Reservoir Exploration and Appraisal

I would like to dedicate this book to the following people:

To my family: My wife Vania and my son Liam, who have always supported me in doing this project and motivated me to finish it.

To my parents, for the example given to me to always look forward.

To my colleagues and coworkers in the oil companies I worked for and those who have provided valuable feedback to me along my career.

Reservoir Exploration and Appraisal

Luiz Amado

AMSTERDAM • BOSTON • HEIDELBERG • LONDON
NEW YORK • OXFORD • PARIS • SAN DIEGO
SAN FRANCISCO • SINGAPORE • SYDNEY • TOKYO
Gulf Professional Publishing is an imprint of Elsevier

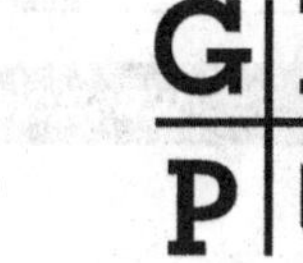

Gulf Professional Publishing is an imprint of Elsevier
The Boulevard, Langford Lane, Kidlington, Oxford, OX5 1GB, UK
225 Wyman Street, Waltham, MA 02451, USA

First published 2013

British Library Cataloguing-in-Publication Data
A catalogue record for this book is available from the British Library

Library of Congress Cataloging-in-Publication Data
A catalog record for this book is available from the Library of Congress

ISBN: 978-1-85617-853-2

For information on all Gulf Professional Publishing publications visit our website at store.elsevier.com

This book has been manufactured using Print On Demand technology. Each copy is produced to order and is limited to black ink. The online version of this book will show color figures where appropriate.

CONTENTS

PREFACE

THE ORIGIN OF THIS BOOK

The idea to write a book on this subject came to my mind for different reasons. First of all because there is a lack in the technical literature about reservoir engineering applied to exploration and appraisal phases of an oil or gas field.

Secondly because during those years during which I have worked as a reservoir engineer providing input to exploration teams, I have gathered much specific knowledge on this subject that I felt would be advantageous to share with others who will probably face in the future the same challenges that I did.

Thirdly because never before have lease sales, bid rounds, farm in/out, data rooms, and partnerships among companies been so to the forefront in the agenda of oil and gas professionals these days, so to understand the fundamentals governing how volumes can be translated to value is mandatory to all, including technical and nontechnical staff.

I believe a great deal of knowledge and experience is needed to screen opportunities in exploration; however, there is a recipe that will reveal how you should approach this and, in the absence of data, how to create a project that will be robust enough to pay off.

I hope this book will be used by you not only as a quick reference but also as a working guide, where you can find the tools which will enable you to perform your job better.

THE CONTENT AND PURPOSE OF THIS BOOK

This book will cover all the steps necessary to evaluate a given accumulation (hydrocarbon volume) and translate it into an economical value in order that it can be included in the portfolio of projects and properly ranked against other similar investment opportunities.

The chapters are structured according to the workflow usually applied by companies to assess the value of their leads and prospects. They will cover the use of analogs, how to derive rock and fluid trends when few or no data at all is made available, and how to build initial development plans and production functions that will feed the final economics of these opportunities.

Special focus will be given to how to derive recovery efficiencies for the wells and reservoirs and how to define the number of producers and injector wells required for a particular field development. The generation of production curves will be explained in detail by using simple and detailed models.

Clastics and carbonates reservoirs will be shown and discussed according to their characteristics and how these rock types will influence the economics in terms of productivity.

The book will also cover not only the exploration but also the appraisal phase as well, the facilities and subsea aspects of those projects and the schedule and development planner work from the first exploratory well until the first oil arrives.

A dedicated chapter on technology is mandatory to quantify how it will improve poor projects and enhance their economics.

THE PEOPLE AT WHOM THE BOOK IS AIMED

Prospect evaluation is a multidisciplinary task and therefore this book is intended for a large audience, namely those in charge of or a member of exploration and appraisal teams and will want to understand how to transform a given in-place volume to a recoverable volume and how to extract economical value of it.

If you are, or work together with, a petroleum engineer, economist, asset manager, geologist, or seismic interpreter, among others, you will certainly benefit from the lessons taught in this book.

If you are a professional in another discipline or industry and want simply to understand how the oil companies acquire and grow their portfolios of projects, you should use this book as a bridge to put you there.

So I hope you will enjoy reading this pioneering book and will apply the methodology described here in your day-to-day work, and let the chapters of this book be your guide to help you create value for your exploratory prospects wherever they may be.

Luiz Amado

INTRODUCTION

Overview of Chapters

This introduction presents the chapters in the order they appear in the book, describing their content and giving the reader an overall view of the logical sequence usually adopted during a prospect evaluation exercise. It was written to serve as a quick reference and guidance to what level of information is available in each of the chapters.

THE YTF NUMBER AS AN ADDITIONAL MOTIVATION FOR READING THIS BOOK

The YTF or "yet to find" number represents the potential of resources that are still to be discovered in a given play or region.

Current discovered reserves are in the order of billions of barrels. However, it is believed that the YTF encompasses over 800 billion barrels around the globe, including areas such as the Arctic and the Sahara desert as well as the North Sea, Gulf of Mexico and Offshore Brazil, and Africa.

Therefore, this huge number gives us a further incentive to understand and implement the topic covered in the book. It will be the use of the techniques and workflow described here that will make it possible to mature the YTF number first in discoveries and later on in reserves.

This book will help you to properly assess and hopefully evaluate this vast number of resources that are still out there and enable them to be extracted economically. Otherwise they will remain unexploited in the subsurface still in the form of a YTF figure.

Chapter 1—Project Framing and the Volume to Value Evaluation Analysis

In the first chapter is explained the methodology used by oil and gas professionals and oil firms in general when evaluating prospects and opportunities and how to create a volume to value workflow. How the project is screened is dealt with along with the main assumptions that should be followed when dealing with any VV type of project.

Chapter 2—Setting the Scene: Plays, Leads, and Prospects

This chapter provides definitions that are important to setting the scene and the objective of this book, that is, the prospect evaluation. Plays and leads are covered as well in order to give an idea of where our asset sits in the context of petroleum exploration.

Chapter 3—Exploration and Appraisal Phases

This chapter is dedicated to the exploration and appraisal activities that will impact the monetization of the prospect such as number of exploratory wells, number and location of appraisal wells, and how to consider the E&A (Exploration and Appraisal) costs in a VV project. Seismic and subsurface studies are also mentioned here and how they should be taken into account.

Chapter 4—Rock and Fluid Estimates

This chapter describes the main fluid and rock properties that will govern volume and productivity, and explains correlations along with how to obtain the parameters that are necessary to our work and how to consider a range of uncertainty for these properties.

The idea is to provide the reader with something he/she can use in his/her daily work without having to spend much time analyzing fluid samples or rock data from cores.

The reader will learn how to use data available in the literature or only a few pieces of available information to derive the correlations required in terms of rock and fluid trends.

Chapter 5—The Search for and Use of Analogs

This is a mandatory chapter about field analogs: how to find field analogs that can support our predictions and validate our studies. Commercial databases are described and how to extract with caution the vast list of information that they make available to all.

Chapter 6—Volumes and Recovery Efficiencies

This chapter covers how to calculate volumes and how to estimate recovery factors. This is one of the main chapters in this book. Methods of estimating recovery factors are described and rules of thumb that can be used in the absence of data.

Chapter 7—Wells and Production Functions

This chapter will build a notional development plan for the project: how to derive the number of wells, recovery per well, and production curves necessary for the economic analysis. Initial rates and production curves will be presented and again how to obtain these curves using simple and more complex approaches.

Chapter 8—Facilities and Subsea Engineering

This chapter introduces facilities and subsea engineering basic design, studies, and associated costs. Type of hosts, tiebacks, floating production units, export pipelines, and manifolds will all be discussed here. How to select and how to consider the right facility and evacuation routes will be described.

Chapter 9—CAPEX and OPEX Expenditures

This chapter is dedicated to expenditures related to all investments that are made in advance and the operational expenses that will occur from the start up of production onward.

Approximate figures will be given for these expenses as seen in deepwater projects along with what main assumptions should be taken into account when inputting these values in our project. Available databases that contain these figures will be shown and estimation of cost and methodology will be discussed.

Chapter 10—Lease Sales, Bid Rounds and Farm In

This chapter describes the process of lease sales, bid rounds, and farm in opportunities: the objective of data rooms visits, data acquisition, and the cycle that should be followed to estimate the value to bid or to offer to a partner or to refuse an offer.

It also presents royalty rates and reliefs as well the particulars of bid sales in general with some examples of recent ones held in the Gulf of Mexico.

Chapter 11—The Value and Gains of Technology

This is a chapter where the reader can find out how to consider the impact of technology in such evaluations: how to incorporate technology that will be developed in the near future and what benefits it can provide to make the project more robust.

Some technologies that are not yet mature may be, however, considered in some projects as they will be sufficiently mature in the future when the project starts up production. Examples are faster drilling techniques, more sophisticated FPUs, subsea pumping, and methods of increasing recovery efficiencies, among others.

Chapter 12—Field Cases Evaluations

In this chapter, field cases examples are set out to illustrate the steps covered in this book, allowing the reader to see a full project cycle and how decision makers would use the information. Deepwater projects for oil and gas are shown and commented on step by step to clarify the process and to eliminate any doubts that the reader may still have.

CHAPTER 1

Project Framing and the Volume to Value Evaluation Analysis

Volume to value (VV) is the methodology used by the petroleum companies for the economical evaluation and selection of exploratory opportunities. An exploratory opportunity is the anticipation of a given accumulation of oil or gas based on seismic sections and well data. Using these inputs and a certain degree of knowledge, interpretation, and imagination, it is possible to construct a geological model (a volume defined by several areas, contours, and variable thickness) that will represent the structure of the prospect.

Thus, the search for exploratory opportunities will contain a high degree of uncertainty and risk. The models, based on seismic section and nearby well data, will provide a conceptual design but until one exploratory well is drilled no presence of hydrocarbons can be assumed.

In order to acquire prospects, oil and gas companies usually have to take part in a bid round or lease sale when blocks or cells will be offered for a given period of time after paying a signature bonus to acquire them.

These blocks vary by size depending on where in the world they are located. In the Gulf of Mexico (GOM), they will be a square with sides of 3 miles by 3 miles. In other parts of the world, the sides of the blocks can be 10 times bigger, as for example, offshore Brazil. Figure 1.1 provides an overview of lease blocks in Campos Basin (offshore Brazil).

Companies will search for blocks where it is believed that several geological factors have combined together to produce oil or gas accumulation. These factors are presented in Table 1.1.

Volume to value (VV) is therefore a process of identifying a prospect, calculating its volume, and converting it to a monetary value. Figure 1.2 shows the VV workflow.

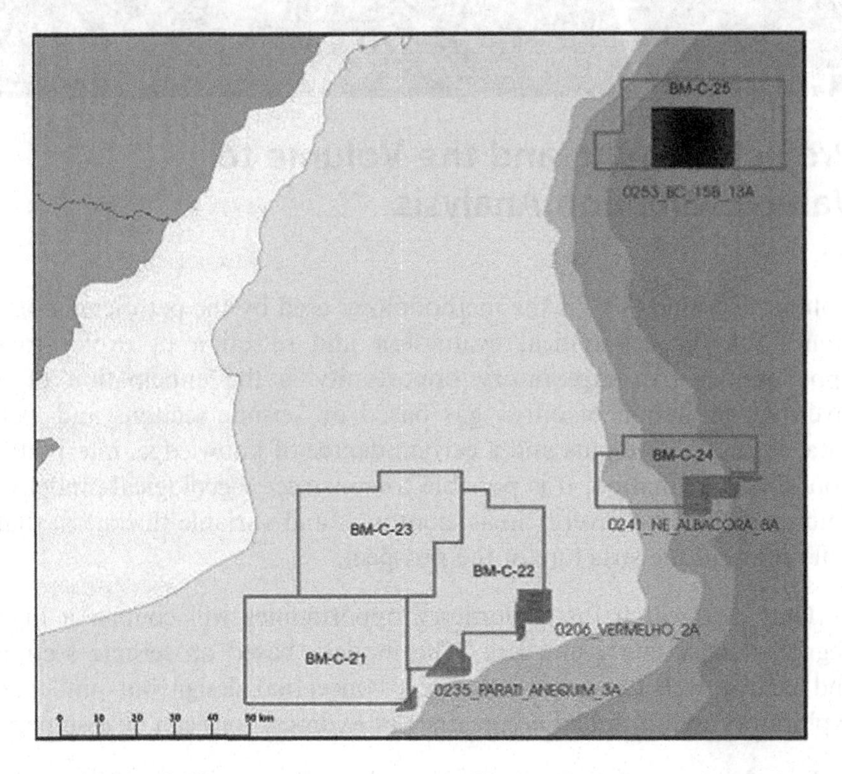

Figure 1.1 Examples of offshore lease blocks (Campos Basin, offshore Brazil). www.anp.gov.br

Table 1.1 Geological Factors

Geological Factors	Definition
Reservoir	Group of sedimentary rocks capable of storing and producing hydrocarbons (oil/gas)
Trap	An arrangement of different features that will concentrate the oil/gas in specific areas
Seal	Rocks that will not allow the oil/gas to escape from the reservoir
Charge	Comprises the source rock and its capabilities to expel and migrate the oil/gas toward the trap

Colors are used for simplification purposes only and have no intrinsic meaning. In simple terms the overall procedure involves the identification of a prospect, followed by estimating its volume, and building a developing plan to recover this volume using wells and facilities, which will demand capital and operational expenditure in order to drill, construct, and place at the field to enable extraction of the resources (volume) over the course of time.

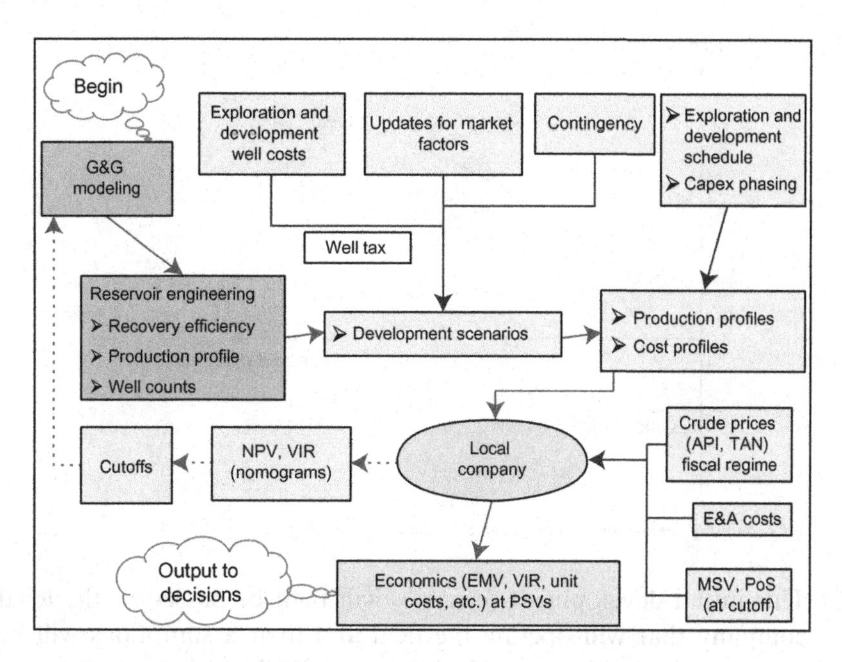

Figure 1.2 Flowchart showing VV workflow.

The volume, once at the surface, will be sold to recover the costs.

The flowchart depicted in Figure 1.2 is self-explanatory. However, let us describe the activities shown there.

1. Geophysics and geology (G&G) modeling—The entire process starts here, with the delineation of the leads and prospects and an estimation of the volumes in place. The volumes will also define the number of exploratory wells needed to be drilled to prove and discover them.
2. Reservoir engineers will then estimate recovery factors (RFs) (according to the rock and fluid properties associated with the accumulation) to apply to those volumes, thereby converting them to recoverable volumes.
3. At the same time they will define the number of development wells (and appraisal wells) and create production profiles that will produce the recoverable volumes with time.
4. Possible development scenarios will be created, each one with their own production and cost profiles. Exploration and appraisal well costs, development well costs, facilities and subsea costs, and operational costs will be estimated and updates on market factors as well as contingency will be added.

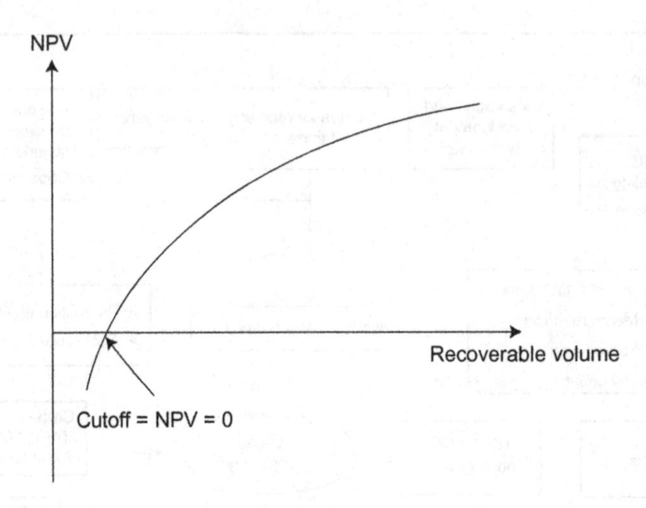

Figure 1.3 Example of nomogram to establish cutoff volumes.

5. This initial development scenario, will then be passed to the local company that will operate the field and their assumptions will be incorporated, such as fiscal regimes, oil and gas prices, and contracts in place as well the tax and royalties due in that particular country.
6. Economics analysis will be assessed and net present value (NPV) and expected monetary value (EMV) figures will be created for all scenarios.
7. A curve, known as a nomogram (NPV plotted against recoverable volumes) will serve as a guide to determine what scenarios are NPV > 0 and those that are NPV < 0 (Figure 1.3).
8. Cutoff volume will then be defined as the intersection on the x-axis that corresponds to NPV = 0.
9. This cutoff will then indicate to the subsurface team the minimal volumes that should be discarded.
10. New mean success volumes (MSV) will be estimated, as we know that we should not use the whole distribution curve of volumes but only that part which produces NPV > 0 projects.
11. The whole cycle is then repeated, the final economics guiding the decision makers in ascertaining what the best approach for the project.

In the next chapter, we will start with the identification and definition of the plays, leads, and prospect.

CHAPTER 2

Setting the Scene: Plays, Leads, and Prospects

In order to understand some of the concepts in this book, we need to define play, lead, and prospect. The definitions given in this chapter are taken from the Internet.

2.1 DEFINITION OF PLAY

In geology, a petroleum play, or simply a play, is a group of *oil or gas fields* or prospects (definition of prospect will follow) in the same region that are controlled by the same set of geological circumstances.

A particular *stratigraphic* or *structural* geologic setting is also often known as a play. For example, in a relatively unexplored area such as the *Falkland Islands*, one might speak of the "Paleozoic play" to refer to the potential oil reserves that might be found within the *Paleozoic* strata.

Another example would be the recent presalt play in Brazil that refers to the strata below the salt layer that was formed before the salt precipitated. In a well-explored basin, such as the *Gulf of Mexico* (GOM), explorationists refer to the "Wilcox play" or the "Norphlet play" to collectively designate the production and possible production from those particular geological formations of *Paleocene* and *Jurassic* age, respectively.

A play may also be a broad category of possible reservoirs or rock types, as in the turbidite play of offshore *Angola* or the carbonate play in the East Java Sea, or refer to the structural geology of the setting, as in the subupthrust play of *Wyoming*.

Sometimes the word play is applied to a geographic area with hydrocarbon potential, for example the South Texas play or the Niger Delta play, but usually "play" is used with the sense of restricting discussion to exploring a particular geological setting.

Thus, one might have both the Wilcox play and the Norphlet play (among others) in partially overlapping areas of the coast of the Gulf of Mexico (GOM); the GOM deepwater play might or might not include elements or particular locations appropriate to either the Wilcox play or the Norphlet play, or to both. The term play may refer to geologic time intervals, rock types, structures, or some combination of them.

2.2 DEFINITION OF LEAD

A lead in *hydrocarbon exploration* is a subsurface *structural* or *stratigraphic* feature with the potential to include entrapped *oil* or *natural gas*. When exploring a new area, or when new data become available in existing acreage, an explorer will carry out an initial screening to identify the possible leads.

Further work is then concentrated on the leads with the intention to mature at least some of them into drillable *prospects* (Figure 2.1).

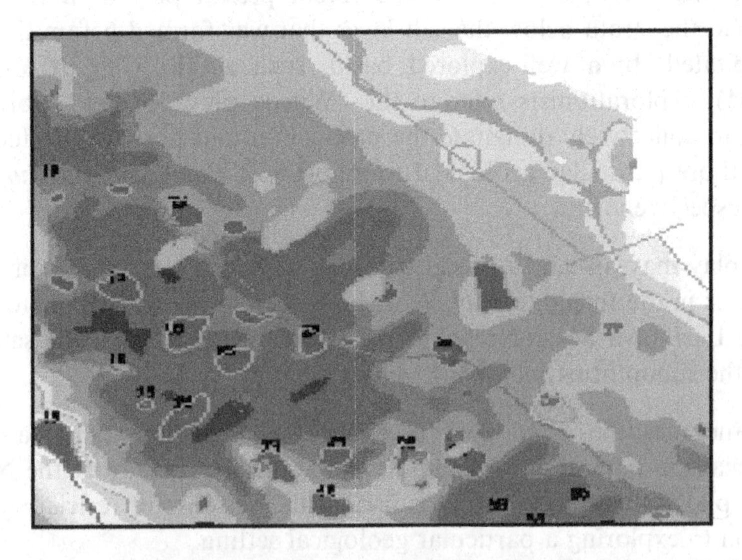

Figure 2.1 Illustration showing possible leads (in magenta) in an offshore basin. (For further interpretation of the references to color in this figure, the reader is referred to the web version of this book.)

A lead is a/an:

1. large simple structure (3 way)
2. ideal charge setting
3. feature with light oil potential
4. presalt source

but *is not a prospect yet.*

1. Understand the results from adjacent wells.
2. Reservoir risk: reservoir presence?
3. Rock quality?

Figure 2.2 Maturing a lead to a prospect.

2.3 DEFINITION OF PROSPECT

A prospect would be an identified volume or accumulation as a result of further work done on a lead in a given play. It is a well-defined structure with a known volume and offering a real possibility for drilling an exploratory well to prove (a) it is there and (b) contains hydrocarbons (Figure 2.2).

CHAPTER 3

Exploration and Appraisal Phases

Once the prospect has been identified and is ready to be drilled, an initial exploratory well is created. This well will discover the accumulation and prove the assumptions made.

The exploration phase comprises the work done by geoscientists to interpret seismic sections, build top and base structure maps, and define the location for an exploratory well and subsequent appraisal wells. Some prospects, owing to their size, may not always require an appraisal well.

3.1 EXPLORATORY WELLS

Usually, a prospect will have one or two exploratory wells, depending on its size and degree of reservoir compartmentalization. Exploratory wells will be planned to test a given volume or area and not the whole prospect. Once the well is a discovery, other wells called appraisal wells will be necessary to delineate the extension of the reservoir and the fluid contacts (not always found in an exploratory well).

On the other hand, in a case where the exploratory well finds no hydrocarbons or misses the reservoir, the prospect will be abandoned. In the past 4–5 years, approximately 40 exploratory wells have been drilled in the deepwater Gulf of Mexico (GOM), with a success rate inferior to 50%, meaning that about half of the wells did not find any commercial volumes of oil or gas (Figures 3.1–3.3).

3.2 APPRAISAL WELLS

To define the number of appraisal wells is a bit more complex. In this case, as the objective is to delineate the reservoir and test other features it becomes not only a function of volume but also structurally complex.

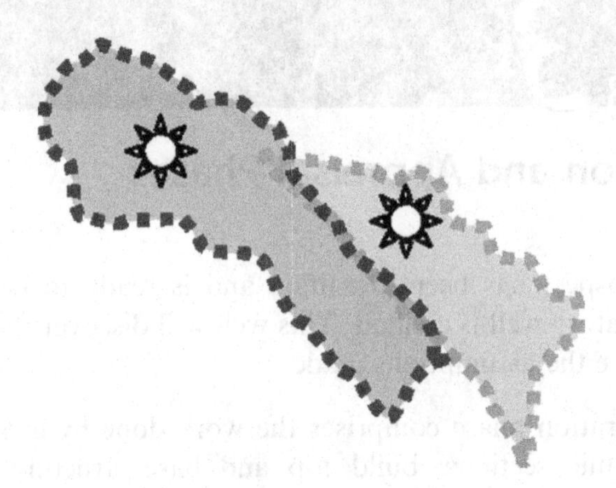

Figure 3.1 Complex structure prospect compartmentalized with two exploratory wells.

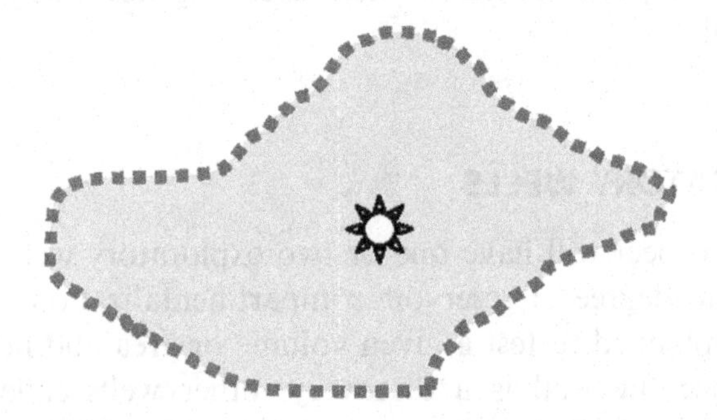

Figure 3.2 Simple structure prospect not compartmentalized with one exploratory well.

General Information	
Geographic area	GOM deep water
Play	Upper Cretaceous
Lease blocks	KC # number
Working interest	50%
Water depth	5000 ft
Main Target	
Age	Lower upper Cretaceous
Trap	Structural (four-way closed)
Stratigraphy	Channel axis
Reservoir depth	20,000 ft
Volume Calculation	
Case	Oil
Recoverable volume	Mean 100, Max 150, Min 50 MM bbl oil
API	30
GOR	500 bbl/scf
Risk Assessment	
Probability of success	21%

Figure 3.3 Example of a prospect summary. GOR: gas-oil ratio; API: American Petroleum Institute–this number defines the quality of the oil (the higher the better); KC: Keathleen Cannon.

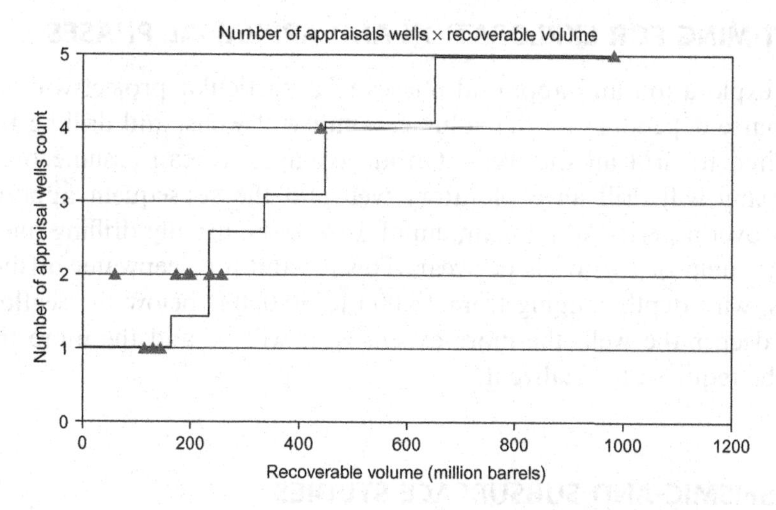

Figure 3.4 Chart to help in defining the number of appraisal wells as a function of recoverable volume. (For further interpretation of the references to color in this figure, the reader is referred to the web version of this book.)

Figure 3.4 provides a quick method of estimating the number of appraisal wells as a function of reservoir recoverable volume (in millions of barrels of oil equivalent). It is supported by field analogs as found in the literature (shown in the figure as blue triangles).

To use the chart, enter the recoverable volume of your prospect and find where it intersects the "staircase" line. The number of wells will be read out on the *Y*-axis. If a fractional number is found, it should be rounded up or down. Remember that the number of appraisal wells will not only be dictated by the volume but also by the complexity of its reservoir structure.

As can be seen, there are a number of prospects for which, for a range of volumes, the number of appraisals is the same. This is due to the fact that a small volume may be too complex or a large volume may have continuity and both will need the same number of appraisals regardless of the size or volume. Note that a maximum number of five is the upper limit as seen in the chart.

As an example, a recoverable volume of 300 MMboe (barrels of oil equivalent) will at first sight require two to three appraisal wells. A gas reservoir of 480 bcf (billions of cubic feet), approximately 80 MMboe, will require one or maybe two appraisal wells depending on its structure.

3.3 TIMING FOR EXPLORATION AND APPRAISAL PHASES

The exploration and appraisal phases of a particular project will vary of course depending on the volume, number of wells, and drilling time required to drill all the wells. On an average, we can assume that a company will drill an exploratory well and the subsequent appraisal wells over a period of a maximum of 3–4 years, usually drilling one or a maximum of two wells per year. This is valid for deepwater offshore wells, with depth ranging from 15,000 to 30,000 ft below the seafloor. The deeper the well, the more expensive it will be and the more time will be required to finalize it.

3.4 SEISMIC AND SUBSURFACE STUDIES

Seismic and subsurface studies costs should be considered in the volume to value (VV) workflow as an exploration and appraisal associated cost. Offshore seismic costs can vary from 3 to 10 MM$.

CHAPTER 4

Rock and Fluid Estimates

4.1 CALCULATION OF OIL AND GAS IN PLACE

Original oil/gas in place is referred to the oil/gas present at discovery. It may be abbreviated as OOIP or OIP for oil, and GOIP or GIP for gas. It should always be referred to as the surface quantity in place. The formulas for calculating it are shown in Table 4.1.

Table 4.1 Formulas for Volume in Place Calculations	
$N = 7758 \cdot A \cdot \phi \cdot h \cdot (1 - S_{wc})/B_o$	$G = 43,560 \cdot A \cdot \phi \cdot h \cdot (1 - S_{wc})/B_g$

The rock properties which will be needed for this calculation are the following: area in acres (A), porosity in fraction (ϕ), and net thickness in feet (h). B_o, B_g (formation volume factors) and S_{wc} (connate or initial water saturation) are fluid properties.

For a prospect, most of these parameters will not be known or will have to be estimated. The methodology described here will combine the use of analogs and correlations.

Areas (in acres) and thicknesses (in feet) will be provided by G&G professionals after interpreting the seismic sections and anomalies seen in their maps. If there is a near well, the thickness may be supported by the one found in the well.

4.2 POROSITY–DEPTH TRENDS

Porosity is obtained from the wells drilled in the area. Porosity versus depth trends usually provide a good source to estimate porosity ranges for different depths in a particular basin, formation, or play.

Figure 4.1 shows two porosity trends with depth for the deepwater Gulf of Mexico (GOM) for two important plays (middle Miocene and Pliocene). As can be seen, this plot can be used (or derived for other

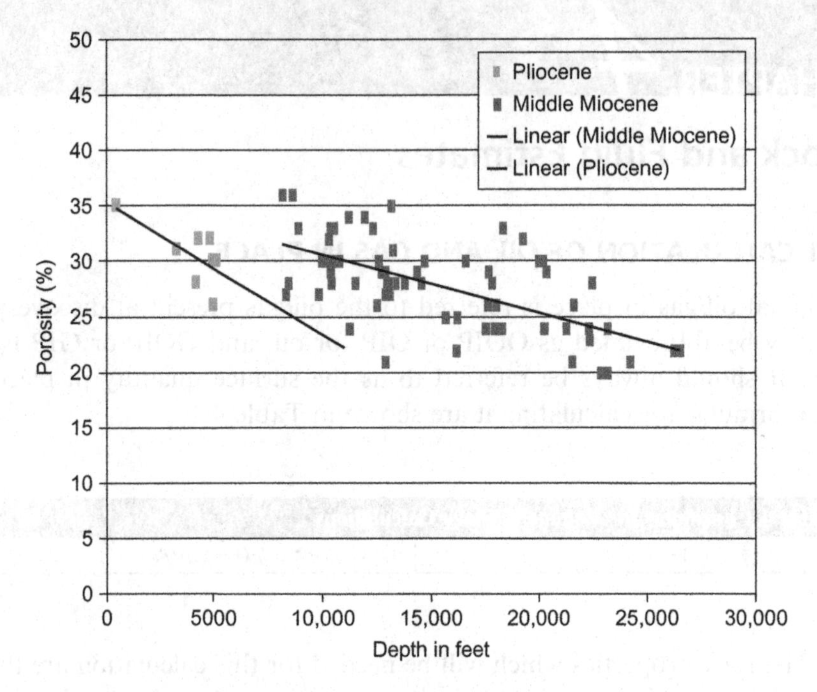

Figure 4.1 Porosity × depth trend for GOM fields in the Middle Miocene and Pliocene ages.

basins or plays) to get ranges of porosity for a prospect situated at a particular depth. Let us suppose our prospect is at 15,000 ft depth. So, we should expect porosities around 28% with ± 5% uncertainty.

4.3 POROSITY–PERMEABILITY TRENDS

Following the same methodology, it is also possible to derive a range of permeability from porosity assuming there is a good trend or agreement between the two parameters.

Figure 4.2 shows for the same Middle Miocene play a correlation or trend for porosity against permeability.

As is expected, the trend increases with higher porosity providing higher permeability. Note that permeability is not required in the volume calculations but will have an impact on the recovery efficiencies and well initials (well rates).

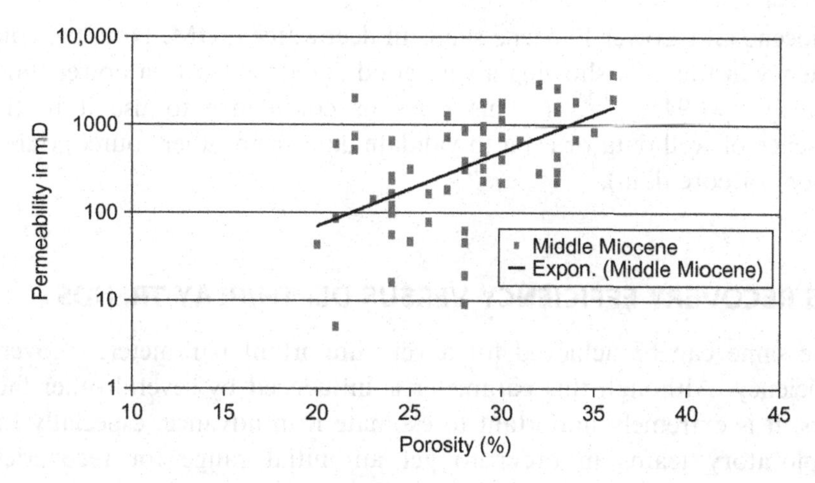

Figure 4.2 Porosity × permeability trend for Middle Miocene fields in the GOM.

4.4 INITIAL WATER SATURATION–PERMEABILITY TRENDS

The next trend is shown in Figure 4.3. It is a correlation between initial water saturation (S_{wi}) and permeability. One should expect that the lower the permeability, the higher the initial water saturation as more water will get trapped by the small pores or network of pores. Again a quick search in the literature or a commercial database or over the Internet will provide us with a good correlation for the Middle

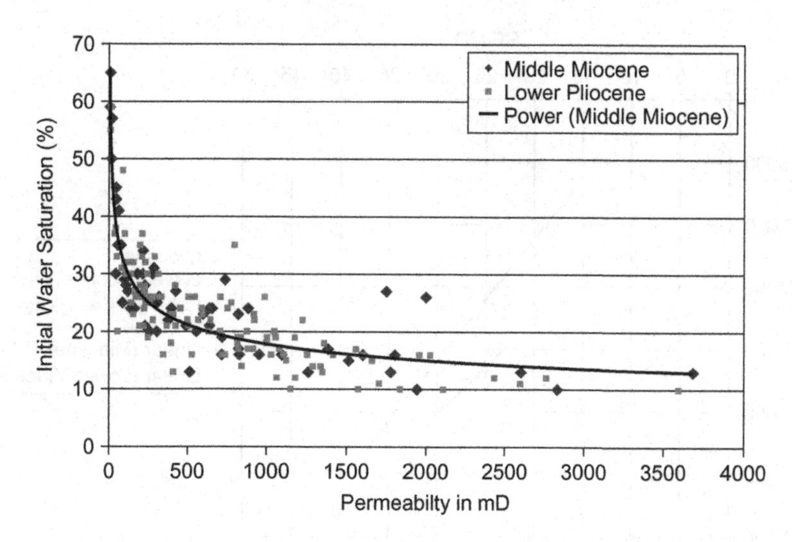

Figure 4.3 Trend showing a good correlation for the permeability and initial water saturations (S_{wi}).

Miocene and Lower Pliocene sands in deepwater GOM. Note the consistency in this plot showing a very good agreement with a power function ($y = 97.945x^{-0.2465}$). This gives us confidence to use it in the absence of well data or even to validate data from other sources (laboratory or core data).

4.5 RECOVERY EFFICIENCY VERSUS DEPTH/PLAY TRENDS

The same can be achieved for a very important parameter: recovery efficiency. Although this parameter is influenced by several other factors, it is extremely important to estimate it in advance, especially for exploratory teams in order to get an initial range for recoveries. Figure 4.4 shows recovery factors (RFs) versus depth for oil and Figure 4.5 for gas fields, for different plays in the deepwater GOM. For oil, it can be seen that recoveries will vary from 10% to 40%.

The Miocene sands seem to show a good trend with depth as does the lower Miocene. The Pliocene sands are the only ones dispersed and not following a particular trend.

This is a good example where more work should be done to group the Pliocene fields according to their reservoir features. Pliocene sands

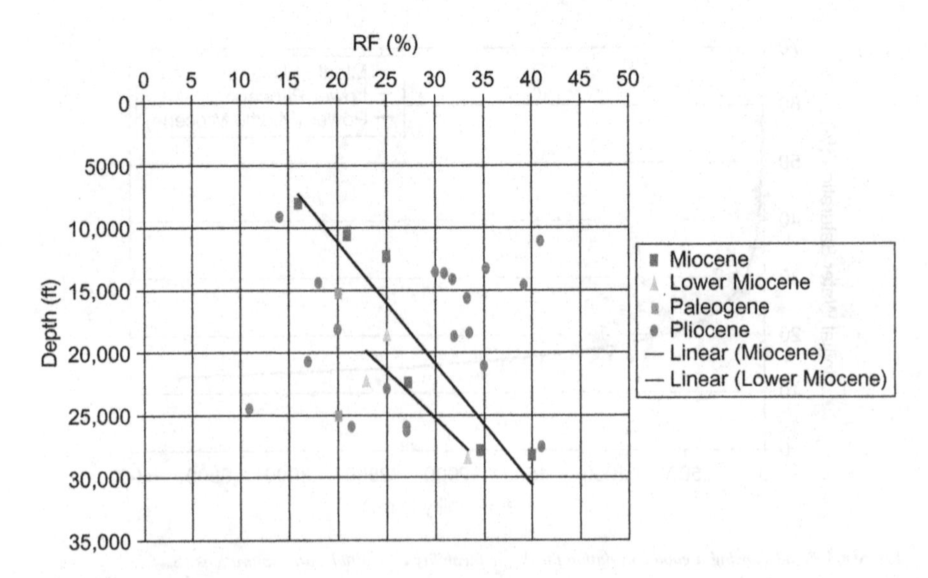

Figure 4.4 Recovery factors (RFs) for deepwater GOM oil fields.

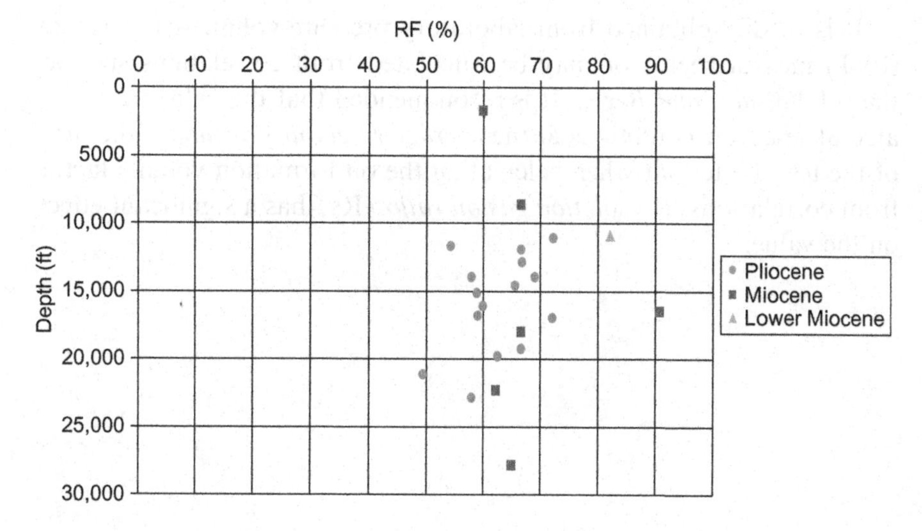

Figure 4.5 Recovery factors (RFs) for deepwater GOM gas fields.

can be set in two groups: one with high permeability and the other with lower permeability.

The group with high permeability presents high recovery (30–40%) and vice-versa (10–25%). Paleogene data shows a lower recovery as expected, due to the heavier aspect of the oil quality, although recently same light oil discoveries have been found in this play (e.g., Tiber prospect).

For gas, the range seen in Figure 4.5 varies from 50% to 90% with an average around 60–70%. Low recoveries may be due to the presence of condensate. No particular trend can be observed. However, the ranges seen in the plot may be used for initial screening of gas recoveries for the plays and depths shown.

4.6 FORMATION VOLUME FACTORS

B_g and B_o can be found using correlations or from laboratory measurements (preferred). It is not the objective of this book to show these correlations, and they are extensively covered in many fluid properties handbooks. The value of the oil formation volume factor is generally between 1 and 2 Rbbl/stbbl (Rm^3/stm^3).

It is readily obtained from laboratory pressure-volume-temperature (PVT) measurements or may be calculated from correlations such as that of *Vasquez and Beggs*. It is recommended that the value be evaluated at reservoir conditions at the *average reservoir pressure* at the time of the test. Note that when calculating the oil formation volume factor from correlations, *the solution gas oil ratio (*Rs*)* has a significant effect on the value.

CHAPTER 5

The Search for and Use of Analogs

To derive a reservoir analog for the Gulf of Mexico (GOM) or any other area is an important task in order to validate/support any evaluation, in particular the numbers that were adopted or used for recovery factor (RF), rock and fluid properties, well initial rates, etc.

It is also mandatory to verify the current portfolio assumptions that were made and check if there are any gaps or inconsistencies. A general workflow is shown in Figure 5.1.

First, pick a good database. It can be a commercial one such as IHS (www.ihs.com) or Reservoir KB (www.reservoirkb.com). Both are available by subscription only. It is also possible to use simply a search tool over the Internet. The next step would be to list all fields in a given area or play (e.g., deepwater GOM or Carbonates in West Africa).

Select those that have potential interest to you. Determine the analog parameters that are relevant to you such as expected ultimate recovery (EUR, well initials, etc.) or a combination of them such as kh/μ.

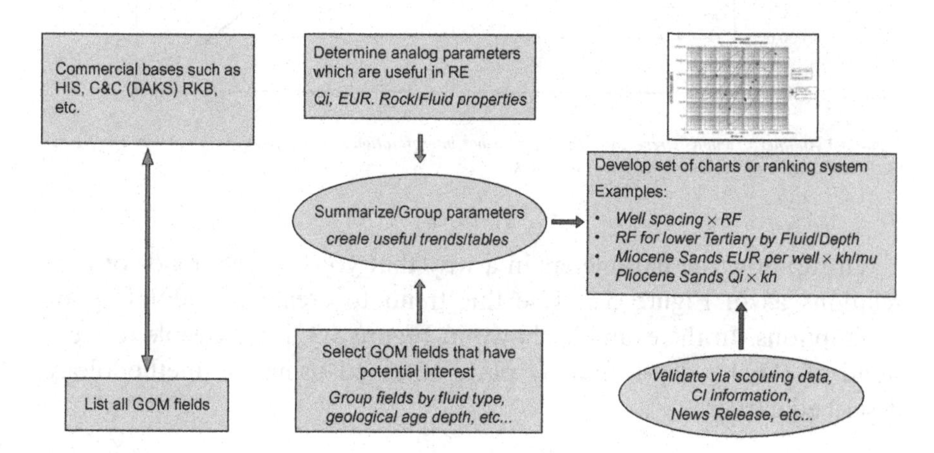

Figure 5.1 A general workflow to build an analog database.

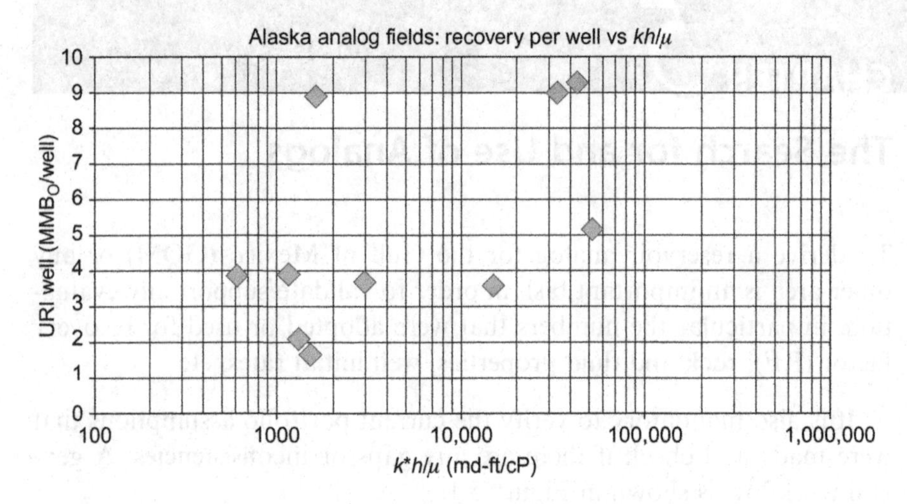

Figure 5.2 An example of a trend deriving using analogs and grouping parameters.

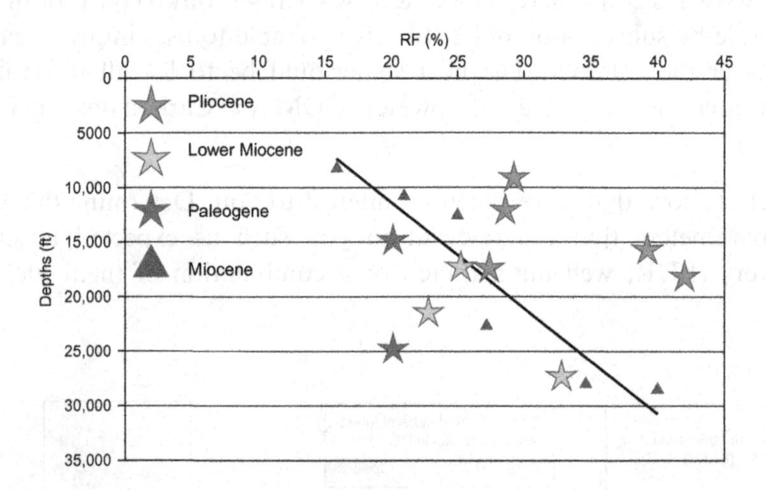

Figure 5.3 Another example—trend for RFs. (For further interpretation of the references to color in this figure, the reader is referred to the web version of this book.)

Group now the parameters in a way that you can see trends or correlations as in Figure 5.2. Use this trend to create or validate your assumptions. In the example shown in Figure 5.3, it is possible to see a trend of RF by depth and by play, obtained using the methodology described earlier.

Table 5.1 Output of a Commercial Database (Edin, from IHS) with Analog Information

Field Name	HC Type	Old Age	Young Age	Top Depth Feet	Oil Recovery PP Factor
Field A	Oil, gas, cond		Upper Pliocene	16,585	75.76
Field B	Oil, gas, cond		Upper Pliocene	16,918	75.76
Field C	Oil, gas, cond		Upper Pliocene	17,561	75.76
Field D	Oil, gas, cond		Upper Pliocene	15,750	75.76

In the same figure, the red triangles represent oil fields in the Miocene play, the orange stars are oil fields in the Pliocene, blue stars are oil fields in the Lower Miocene, and the green stars are oil fields in the Lower Tertiary (Paleogene).

By simply observing the data, it is clearly seen that for the same depth the lowest RF will be found in the Lower Tertiary and the highest in the Pliocene sands. Therefore, we can use this trend to estimate our RFs in these plays, but very carefully. We can also create other trends such as recovery versus oil quality or versus a group of fluid and rock properties. Same databases will contain more geological data. Others will have more engineering data and still others will mix both geological and engineering data with costs (Opex and Capex).

The example in Table 5.1 shows the output of a commercial database in the format of a downloaded spreadsheet. The user can select the information needed and use the information to determine their own trends and correlations.

Other databases will display information as a .pdf file with summary sheet and text and figures, including production profiles, well logs, and geological maps. Recent databases have a graphical interface with maps and automatic plots. It is not the intent of this book to recommend any database. They are cited here as reference only. The general idea is to give the reader of how to start a simple search of analogs and how to use the information obtained in the most productive way for the project.

5.1 USEFUL TRENDS DERIVED USING ANALOGS

Figures 5.4–5.7 contain make up data and are shown solely as an example of properties trends.

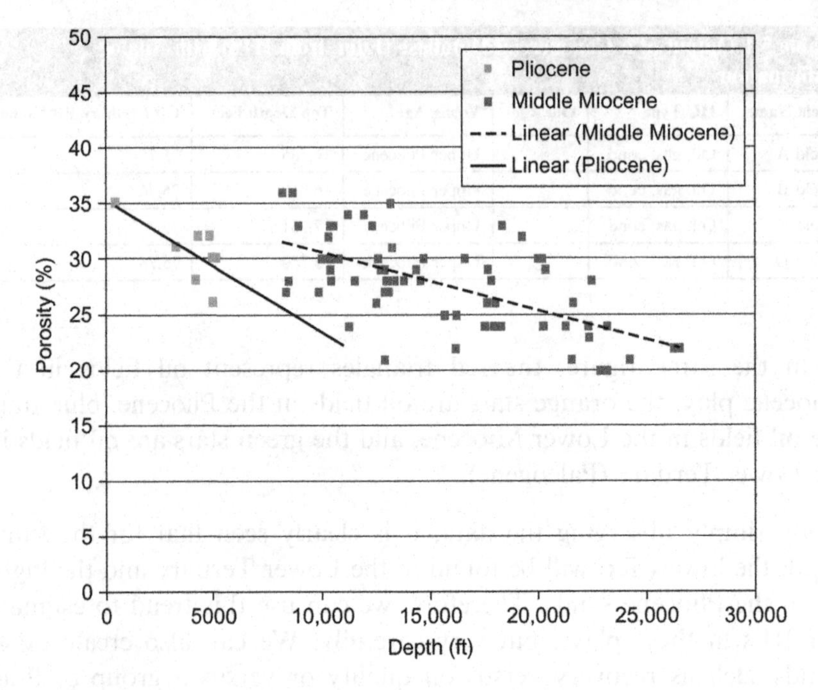

Figure 5.4 Example of porosity trend with depth.

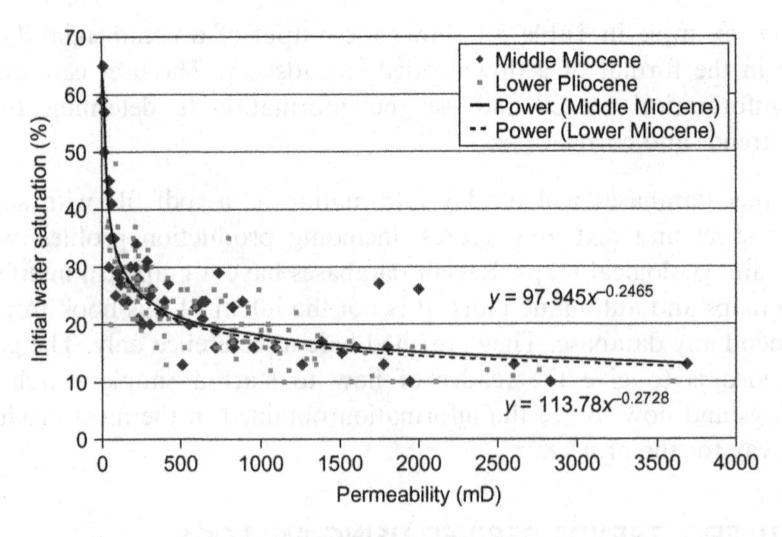

Figure 5.5 Example of water saturation trend with permeability.

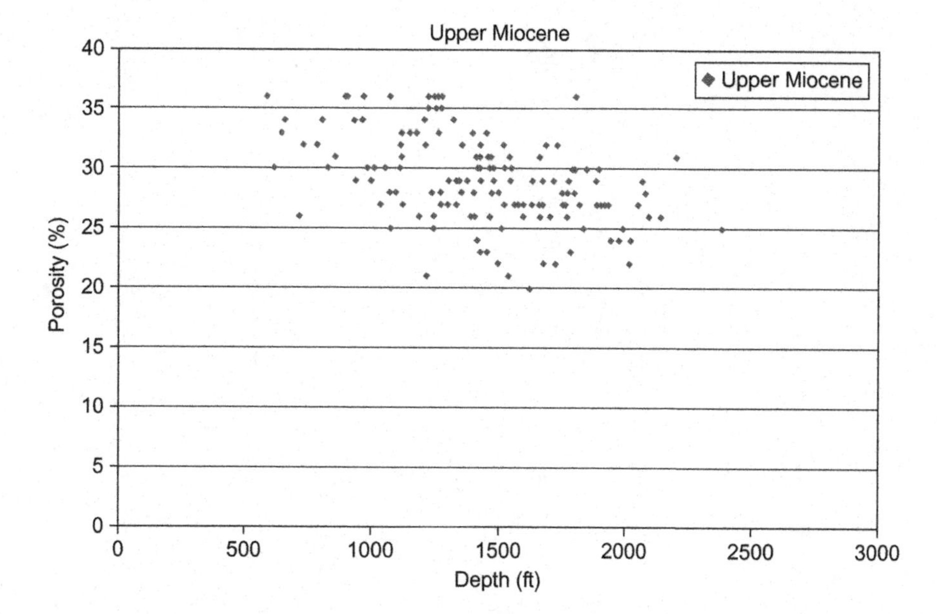

Figure 5.6 Example of Upper Miocene porosity trend with depth.

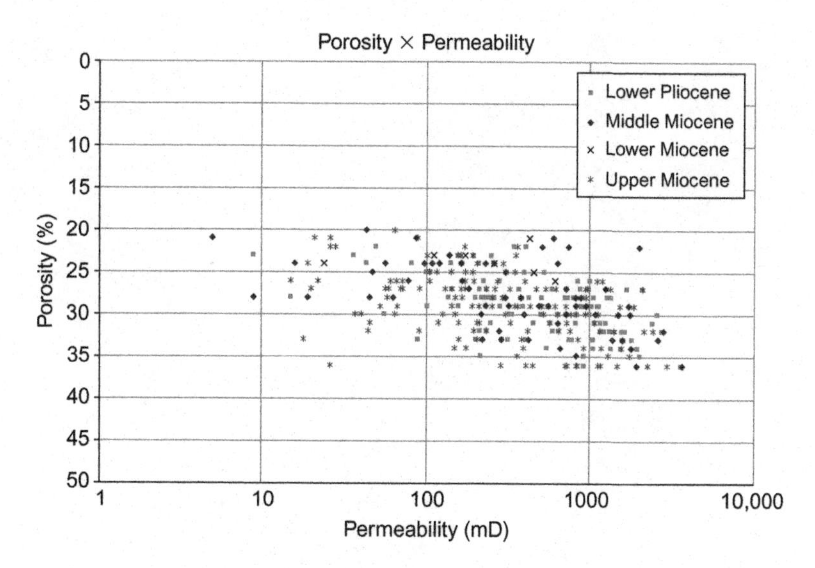

Figure 5.7 Example of porosity trends in other plays with depth.

CHAPTER 6

Volumes and Recovery Efficiencies

In Chapter 4, formulas are given to calculate volumes in place. However, we need to recover that volume and therefore a recovery factor (RF) or recovery efficiency is needed. RF is a function of many parameters. It can be obtained by reservoir simulation or estimated using analogs or analytical methods.

For a prospect, most of these parameters will not be known or will have to be estimated. The methodology described here will combine the use of analogs and correlations as well as the use of analytical models.

6.1 RF TRENDS BY ANALOGS

Figure 6.1 shows an RF trend with depth for important deepwater Gulf of Mexico (GOM) plays (age). It is the same plot as that shown in Figure 5.3 but with trend lines or regression lines that have been derived from the data.

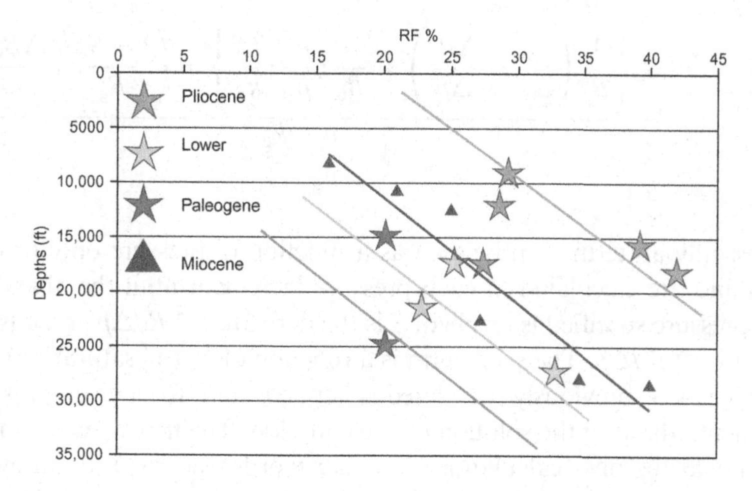

Figure 6.1 RFs versus depth after fitting straight lines.

As can be seen, this plot can be used (or derived from other basins or plays) to get ranges of recoveries for a prospect situated at a particular depth and/or age.

However, certain other items of information can be used to validate this plot or at least provide a bit more confidence in the trends shown. Analytical solutions is one such.

6.2 EUR/WELL AND RF TRENDS BY ANALYTICAL SOLUTIONS

The radial form of Darcy's equation, for example, is a simple equation that can be used for validation purposes at this earlier phase if combined with an appropriate decline model to predict how productivity will drop with time.

$$q = 0.00708kh(p_e - p_w)/(\mu B \ln(r_e/r_w)) - \text{Darcy's model}$$

$$\text{EUR} = q_i \cdot k \cdot e^{-a(t)} - \text{Decline model}$$

with a being the annual decline rate (a can be input as a distribution).

For RFs, there is a similar procedure described in an old SPE paper (SPE 000469-G, from J.J. Arps and others) where the authors derive a depletion and an RF equation.

The depletion equation is shown below, and it can be solved numerically in small pressure drop increments. The paper describes the process in detail and the assumptions made.

$$\frac{\Delta S_o}{\Delta p} = \frac{S_o \left\{ \frac{1}{B_g}\left(\frac{s}{B_o} - \frac{\Delta B_g}{\Delta p}\right) + \frac{b}{B_o} \cdot \frac{\mu_o}{\mu_g} \cdot \frac{k_g}{k_o} \right\} + \left(\frac{1 - S_w}{B_g}\right)\frac{\Delta B_g}{\Delta p}}{1 + \frac{\mu_o}{\mu_g} \cdot \frac{k_g}{k_o}}$$

Basically all terms can be seen as a function of pressure only (except k_g/k_o) and are calculated in each pressure increment until the abandonment pressure specified is reached. S is the derivative $\Delta R_s/\Delta p$ and b is the derivative $\Delta B_o/\Delta p$. The k_g/k_o term is a function of S_o (oil saturation) and the value of S_o previously calculated is kept constant for a given pressure increment, allowing the solution of the equation. The new S_o is then kept constant in the next calculation. In other words, the equation above is solved explicitly for S_o and implicitly for pressure. Once we have the final

S_o at the abandonment pressure, the RF can be calculated again according to the SPE 000469-G paper simply on the basis of the difference between S_o final and S_o initial and the respective B_o values (formation volume factors).

$$\text{RF} = 77.58\left(\frac{S_o}{B_o} - \frac{S_o'}{B_o'}\right)\text{STB/acre ft/percent porosity}$$

The analytical modeling will also be improved with the use of a probabilistic range of "decline rates" to simulate, for example, EUR/well or the depletion case equation. The range for the parameters in the previous formulas are usually based on a particular play or rock-type field performance and/or simulation modeling.

In the case of Darcy's radial flow equation, the analytical model for "decline rate" (to estimate the $p_e - p_w$ term with time) is used as a general proxy for subsurface reservoir influences:

- operating conditions (drawdown, artificial lift, and water injection);
- production well interference;
- faulting/baffles which prohibit pressure communication;
- reservoir vertical heterogeneity (i.e., poor sweep/accelerated water breakthrough);
- relative permeability effects.

6.3 PROCESS ADOPTED

1. Use existing "agreed input distributions" for the basis of modeling work.
2. Aggregate input ranges to ensure broad enough capture for all prospects.
3. Combine a single-well inflow model and decline model to determine UR/Well estimate or a $\Delta S_o/\Delta P$ estimate.
4. Apply Monte Carlo or @RISK analysis to determine the range of possible outcomes.
5. Cross plot kh/μ versus UR/Well using Monte Carlo outcomes.

6.4 INPUT DISTRIBUTIONS

Table 6.1 provides a summary of an input table with input distributions that can be considered as input for the Monte Carlo analysis.

Table 6.1 Variables and Distributions

Description	Min	P90	P50	Mean	P10	Max	SD	Distribution Type
Gross thickness (ft)	82	–	–	754	–	1312	300	Truncated log normal
NTG (fraction)	0.12	–	–	0.40	–	0.80	0.175	Truncated log normal
Permeability (mD)	–	7	38	60	256	–	210	Truncated log normal
Porosity (fraction)	0.06	–	–	0.11	–	0.26	0.05	Truncated normal
Water saturation (fraction)	0.15	–	–	0.25	–	0.35	0.1	Truncated normal
Formation volume factor (rb/stb)	1.20	–	–	1.40	–	1.90	0.3	Truncated normal
Fluid viscosity (cP)	0.30	–	–	1.20	–	4.00	0.5	Truncated normal
Skin (dimensionless)	−2.00	–	–	0.00	–	1.00	–	Triangular
Vertical drawdown (psi)	1000	–	–	1500	–	2500	500	Truncated log normal
Decline properties	0.10	–	–	0.150	–	0.40	0.1	Truncated normal

6.5 EXAMPLE USING @RISK PROGRAM

Table 6.2 provides a listing for an example of how to structure the input distributions in an Excel spreadsheet and use the add-ins @RISK to generate the distributions and run the iterations. In this example, the EUR/Well versus kh/μ is calculated using the equations described a priori.

Table 6.2 Example with Input and Results

Probabilistic Inputs (all others taken as constant)							
Rock properties	**Min**	**P90**	**P50**	**Mean**	**P10**	**Max**	**Distribution type**
Gross thickness (ft)	82			377		656	Log normal
NTG (fraction)	0.12			0.50		0.80	Log normal
Permeability (mD corrected for oil)	4	7	38		256	640	Log normal
Fluid properties							
Formation volume factor (rb/stb)	1.05			1.30		1.50	Normal
Fluid viscoity (cP)	0.50			0.83		1.50	Normal

(Continued)

Table 6.2 (Continued)							
Probabilistic Inputs (all others taken as constant)							
Rock properties	**Min**	**P90**	**P50**	**Mean**	**P10**	**Max**	**Distribution type**
Inflow properties							
Skin	−0.50			1.00		3.00	Triangular
Vertical drawdown	1500			2000		2500	Log normal
Horizontal drawdown	1000			1500		2000	Log normal
Decline properties	0.20			0.25		0.30	Normal
Darcy inflow model							
k—Average permeability (to primary phase)	112	mD	Log normal distribution				
h_g—Gross formation thickness	377	ft	Log normal distribution				
NTG—Net to gross ratio	0.37	fraction	Log normal distribution				
μ_o—Oil viscosity	0.83	cP	Normal distribution				
B_o—Formation volume factor	1.30	rb/stb	Normal distribution				
r_e—Drainage radius	2106	ft	Single value				
r_w—Wellbore radius	0.354	ft	Single value				
s—Skin	1.2		Triangular distribution				
Transmissibility/viscosity (kh/μ)	**18,683**	**mD/cP**	formula kh/μ				
Productivity index	**11.2**	**stb/d/psi**	**Formula Darcy's equation**				
Drawdown	1859	psi					
Initial production rate (at 90% uptime)	**18,737**	**stb/d**	**Normal distribution**				
A—Annual decline rates (formula $Q_i \cdot k \cdot e^{-3(t)}$)	0.25	fraction	Normal distribution				
Y—Years of production	20	years	Normal distribution				
EUR/Well estimate (at 90% uptime)	**28.1**	**MMstb**	Formula decline equation				

6.6 RESULTS

Figure 6.2 summarizes the outcomes from the Monte Carlo simulation after running @RISK for over 5000 or more iterations and plotting for every kh/μ the range of EUR/Well found. Similar plots can also be done for depletion and RF equations.

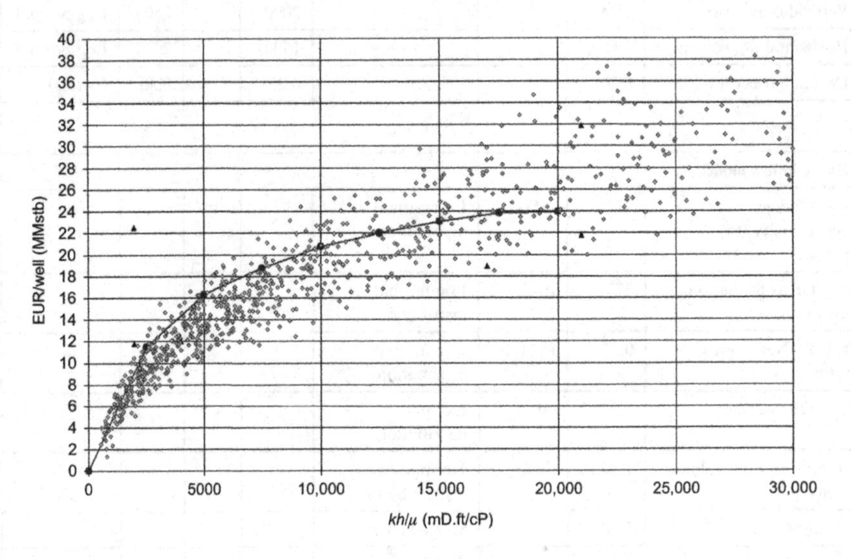

Figure 6.2 Results showing EUR/Well versus kh/μ.

CHAPTER 7

Wells and Production Functions

This chapter will build a notional development plan for the project: how to derive the number of wells, recovery per well, and production curves necessary for the economic analysis. Initial rates and production curves will be presented and again how to obtain these curves using simple and more complex approaches.

7.1 WELL COUNT ESTIMATION AND PRODUCTION CURVES

To create a notional development plan for a field, we need to estimate basically the number of wells and, afterward, their initial rate. It will represent the relationship to rock and fluid properties. In this book, we will use an empirical approach that can be supported by analogs. Figure 7.1 shows the relationship between *kh* and μ with EUR/well. The dots represent fields worldwide where we were able to extract information. Names are not shown to maintain confidentiality.

Provided that we get an idea of the permeability, net thickness, and fluid viscosity, we can estimate the EUR/well. This provides a quick means of estimation, and it is not intended to replace full field modeling, simulation, or any other more robust technique.

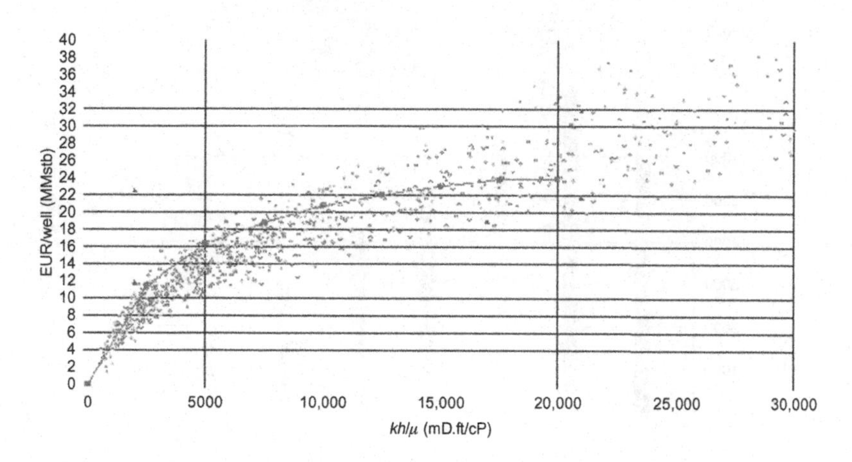

Figure 7.1 EUR/well as a function of kh/μ.

The reader can create their own chart, adding other parameters or trends that they think more appropriate.

Now that we have the number of wells, we can create a production curve. Third-party software is available to provide material balance calculation and decline curves analysis to generate production profiles.

The profiles can also be generated assuming a decline rate based on analog fields and scaled up and down in order to match the recovery (volume) we have. Another quick way is to use tank models or material balance equations as described in reservoir engineering and production handbooks. In the absence of any analog, a quick well profile can be derived using the decline equation:

$$Q/Q_i = (1 + B_i \times a \times \Delta T)^{-1/a}$$

where:

Q_i = initial rate
B_i = initial field decline
a = hyperbolic decline exponent
ΔT = time $(T - T_i)$

The example in Figure 7.2 shows the application of the decline equation with Q_i = 10,000 bbl/d, a = 1.5, and B_i = −0.15 to create the production curve listed in Table 7.1.

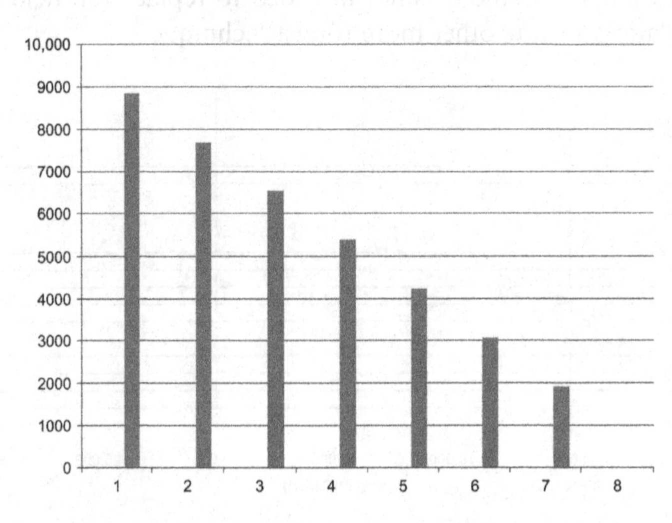

Figure 7.2 Example of generation of a quick production profile using the decline equation.

Table 7.1 Parameters Used to Generate Decline and Production Profiles	
T	$Q/Q_i = (1 + a \times B_i \times \Delta T)^{-1/a}$
0	
1	8844.811482
2	7689.622964
3	6534.434447
4	5379.245929
5	4224.057411
6	3068.868893
7	1913.680375
$Q_i = 10{,}000$, $a = 1.5$, $b = -0.15$	

CHAPTER 8

Facilities and Subsea Engineering

This chapter introduces facilities and subsea engineering basic design, studies, and associated costs. Types of hosts, tiebacks, floating production units, export pipelines, and manifolds will all be discussed here. How to select and how to consider the right facility and evacuation routes will be described. Offshore deepwater will be dealt with: floating production storage and off loading system (FPSO), floating production system, usually in circular shape (SPARs), semisubmersible, and tension leg platform (TLP) as a standalone host. These will offer a solution in remote areas where no subsea tieback is feasible. Subsea tieback will afford a solution where we have an extant facility situated 30 miles or so from the field. In such a case, the subsea (wells and manifolds) can be connected to a host (Figures 8.1 and 8.2).

Although it may be possible to tieback fields situated at a distance greater than 30 miles, flow assurance studies will be necessary to guarantee the feasibility of long tiebacks of up to 90 miles.

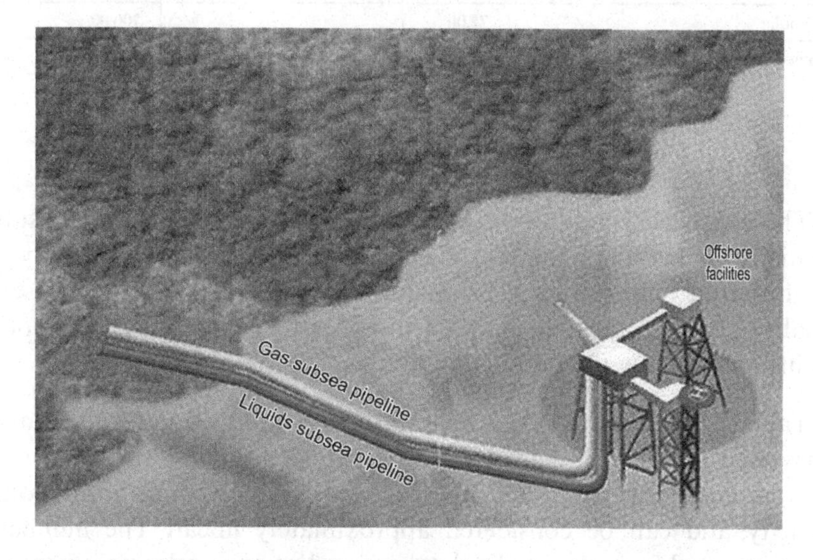

Figure 8.1 Illustration with an example of a hub or standalone platform.

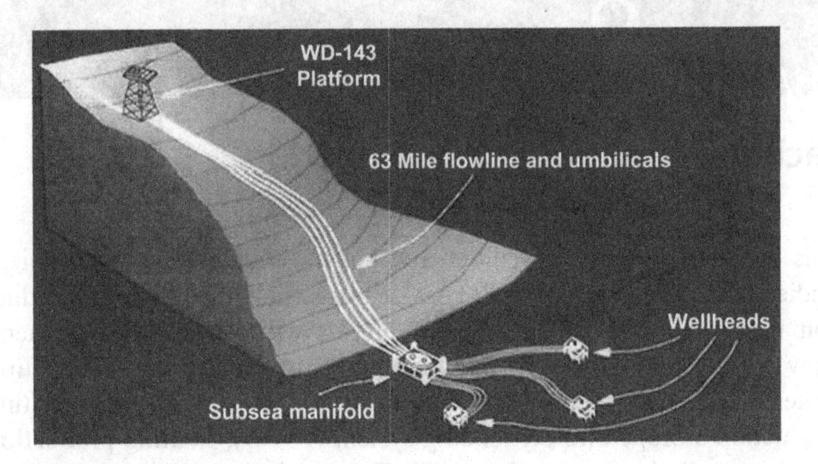

Figure 8.2 Illustration of a subsea tieback.

Table 8.1 Facility Capital costs Versus Facility Peak Rates for Main Gulf of Mexico Deepwater Fields

Field (Facility)	Peak Rate (Mbopd)	CAPEX (MMUS$)
Ram Powel TLP	70.00	1000.00
Blind Faith	45.00	900.00
Mad Dog (SPAR)	100.00	600.00
Jack St Malo	120.00	1500.00
Tahiti FPU	135.00	1800.00
URSA TLP	150.00	942.50
Magnolia	75.00	700.00
Thunder Horse	250.00	5000.00

The capital expenditure (CAPEX) involved for a hub will be substantially higher than that for a subsea tieback, where no facility is required, but only subsea wells, manifolds, and pipelines. Tieback will usually require a tariff to be paid in order to use another operator's facility to evacuate the production.

Table 8.1 shows CAPEX expenditures for different facilities and types used in known deepwater Gulf of Mexico (GOM) fields. As can be seen, costs vary from 600 to 1800 MMUS$, as a function of total capacity and can be considered approximately linear. The numbers were extracted from the public database and/or press releases.

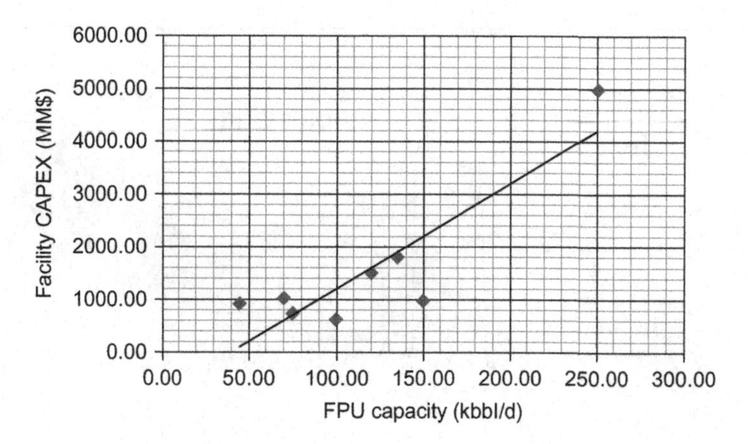

Figure 8.3 Facility CAPEX versus facility capacity for the main GOM fields.

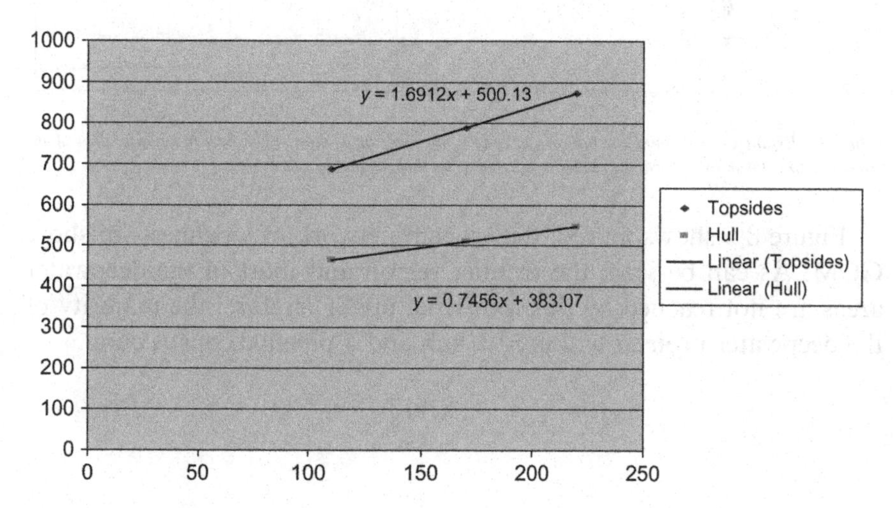

Figure 8.4 Facility costs subdivided into hull costs and topsides costs.

Figures 8.3 and 8.4 provide us with values for facilities' CAPEX and can be used as guidelines for new projects as most of the developments will be in the ranges of rates shown. Likewise we will now need to estimate subsea and pipelines costs. These costs are based on the number of wells (manifolds) and length and diameter (of pipelines) and will be covered in Chapter 9.

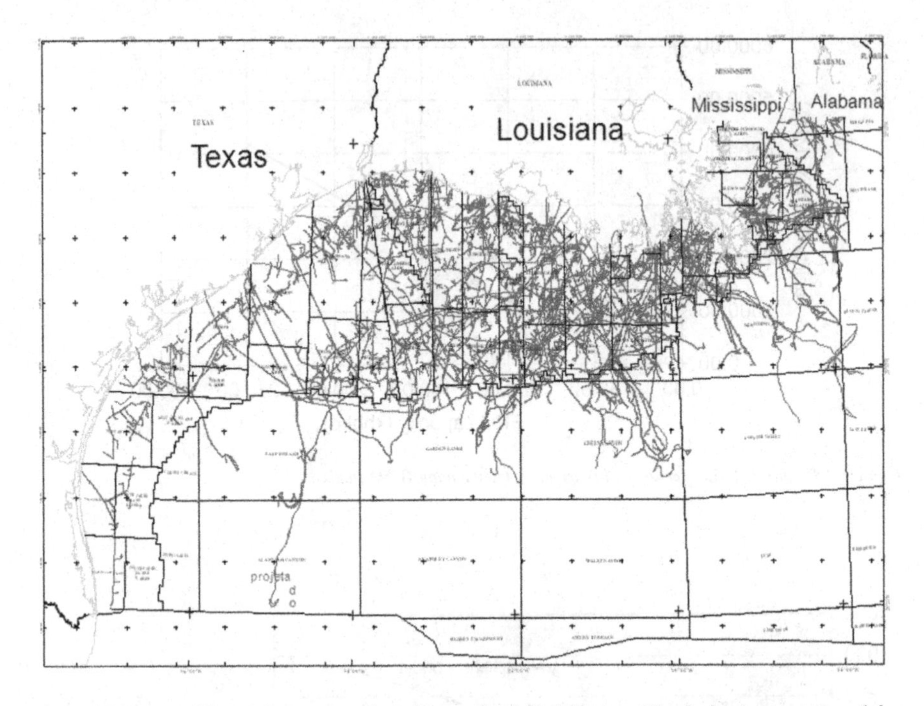

Figure 8.5 Extant offshore infrastructure of pipelines in the Gulf of Mexico. (For further interpretation of the references to color in this figure, the reader is referred to the web version of this book.)

Figure 8.5 shows, in red, the existent network of pipelines—offshore GOM. As can be seen, the frontier region and most of the deepwater areas are not reached by this infrastructure. Therefore, the majority of the deepwater projects will need a hub and a pipeline construction.

CHAPTER 9

CAPEX and OPEX Expenditures

This chapter is dedicated to expenditures related to all investment that is made in advance, and the operational expenses that will occur from the start up of production onward. Approximate figures will be given for these expenditures as seen in deepwater projects and we will look at what main assumptions should be taken into account when inputting these values to our project. Available databases that contain these figures will be mentioned and estimation of cost and methodology will also be discussed.

9.1 FACILITIES AND SUBSEA COSTS (CAPEX)

CAPEX can be defined as the costs for drilling the wells and constructing the necessary facilities (offshore platforms, FPSOs, SPARs, semi-submersibles), subsea manifolds, and pipelines.

The integrated or independent oil and gas companies (IOC) have their own benchmarks and internal costs for all these expenditures. This book intends to show approximate numbers in order that the reader can estimate an order of magnitude for running projects, and benchmark with your own costs.

9.2 DRILLING COSTS

The magnitude of such costs will depend on the depth of the well and the daily rig rate. The daily rig rate will vary according to the rig type, water depth, distance from shore, and drilling depth.

For onshore land, these costs will total <100,000 $/d, and for deepwater offshore Gulf of Mexico (GOM) they can be very high—up to 600,000–800,000 $/d (values for 2010). The number of days will be a function of depth. For usual depths up to 20,000 ft, we can assume 70–80 days and for greater depths up to 32,000 ft, a maximum of 150 days. We can then create the data given in Table 9.1 to estimate the dry hole cost.

Table 9.1 Drilling Costs as a Function of Rig Rate (2010) and Drilling Time

Rig Rate (MM$/d)	Depth (ft)	Drilling Days	Well Cost (MM$)
500–800	20,000	70–80	35–64
500–800	26,000	110	55–88
500–800	32,000	150	75–120

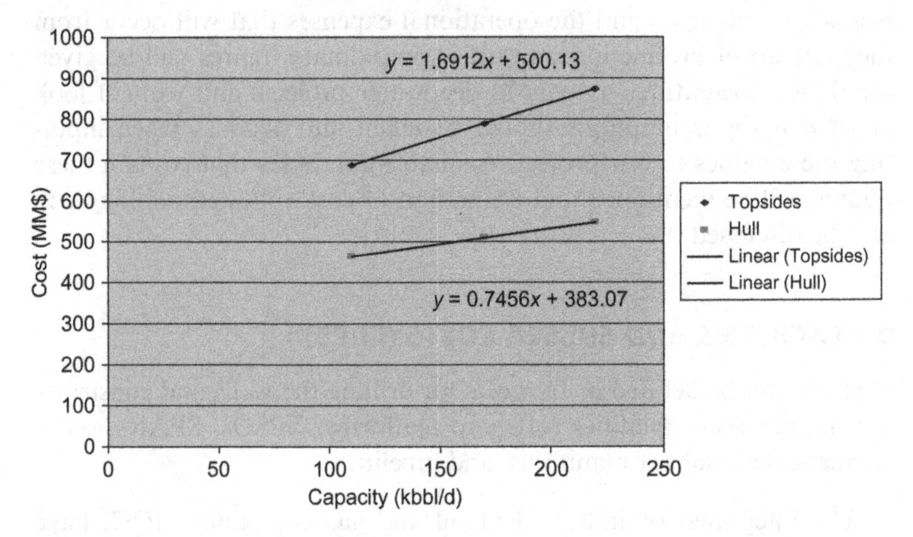

Figure 9.1 Hull and topsides CAPEX estimation versus FPU capacity.

This is the cost of the exploratory well. For appraisal and development wells, we will need to add some extra costs such as evaluation and completion costs. Our volume is 125 MMbbl of oil (so we will need 1 exploratory well + 1 appraisal well, and 13 development wells). One exploratory well will cost 55 − 88 MM$; 1 appraisal well at 55 − 88 MM$ + evaluation costs (15 days extra) = 62.5 − 100 MM$.

Figure 9.1 shows that topsides and hull for semisubmersible production units (SEMIs) may be seen as a function of total capacity and can be considered approximately linear. The numbers were extracted from the literature and press releases and may not be absolutely precise.

The *X*-axis in Figure 9.1 represents unit capacity in thousands of barrels per day and the *Y*-axis the cost (fabrication + installation) in MM$. We want our unit to be sized to accommodate 30 kbbl/d. Using the equations as shown in the plot, we get for 30 kbbl/d, hull cost

~400 MM$ and topside ~552 MM$, with a total =958 MM$. We may want to add some contingency costs here around +25% = 1278 MM$.

The next step is to calculate the subsea costs and export pipelines costs. For 13 wells, we will need two manifolds (one with 6 wells and another with 7 wells) plus interfield flow lines and risers. These costs are usually estimated using figures from a previous project and/or from vendors. Let us make some assumptions here to estimate these costs.

- Well templates and manifolds = 34 MM$/well
- Production riser = 2 risers per manifold = 4 risers at 0.35 MM$/(″ × 1000 ft)
- Export riser = 1 at 0.28 MM$/(″ × 1000 ft)
- Flow lines (FL) and export pipelines at 0.23 and 0.16 MM$/(″ × miles). Let us assume flow lines length = 5 miles and export pipeline length = 30 miles.

These are estimated factors as provided by vendors and publications. Templates and manifolds = 34 × 13 wells = 442 MM$, production risers = 4 × 8 × water depth (1.5 × 5500 ft) × 0.35/1000 = 92.4 MM$, export riser = 0.28 × 5500 ft/1000 ft = 1.54 MM$, FL + export pipelines = 0.23 × 5 miles + 0.16 × 30 miles = 5.95 MM$.

9.3 OPERATING COSTS (OPEX)

The operating costs are more complex and difficult to estimate using only charts or factors. They are usually a function of the peak production rate and tariff rates paid to export production via third-party pipelines.

Table 9.2 offers a tentative summary of OPEX numbers per barrel of oil equivalent for the years applicable on plateau or peak rates based on what has been published by operators in the GOM.

Table 9.2 Total OPEX Estimation

Facility Type	No Tariff	With Tariff
Standalone FPU	8–15 $/boe	12–25 $/boe
Tiebacks	5–10 $/boe	8–13 $/boe

So for this project (standalone with tariff), we will estimate our total OPEX for field life as a minimum of 125 MMbbl × 12 and maximum of 125 MMbbl × 25 = 1500 and 3125 MM$ for the 9 years of production on plateau. This total cost should be prorated proportionally to production for the remaining years.

For the peak rates on plateau, our yearly OPEX would be 1500/9 or 3125/9 and prorated for the remaining years. A minimum of 25% OPEX is recommended for the years where the production is lower than 25% of peak to take into account the maintenance costs and other costs that are not a direct function of the production.

9.4 PROJECT TIMING AND INPUT TO ECONOMIC ANALYSIS

We have now calculated most of the input we need for our economic analysis and to the end of our volume to value (VV) process.

The timing of the project is crucial in relation to all the other inputs, but it is also flexible depending on how aggressive the company wants to be regarding drilling the wells and investing in advance for the construction of facilities.

Ideally, we will drill at least one well per year and allocate a period of time in which to fabricate the facilities required and have them in place. Table 9.3 provides an overview for the total project time from exploration to first production. In the best case scenario this period would not be less than 5 years. For a big project, it can extend to 10 or more years from drilling the exploratory well to first production.

Table 9.3 Timing Period from Exploratory Well to First Production

Facility Type	Time (Years)
Standalone FPU	1 year—Exploratory well, 2 years—Appraisal well 3–7 years (facilities + subsea)
Tiebacks	As above Up to 3 years for the facility tieback

CHAPTER 10

Lease Sales, Bid Rounds, and Farm In

This chapter describes the process of dealing with lease sales, bid rounds, and farm in opportunities. The objective is to appreciate the data acquisition procedure and the overall sequence of events that should be followed in order to estimate the value to bid, offer to a partner, or the basis on which to refuse an offer.

Also presented are royalty rates and reliefs as well as the details of bid sales in general with some examples of recent ones held in the Gulf of Mexico (GOM).

10.1 DEFINITIONS FOR LEASE SALES, BIDS, AND FARM IN OPPORTUNITIES

According to oilfield glossaries, a *lease sale, bid round,* or *licensing round* can be defined as an occasion when a governmental body offers exploration acreage for leasing to exploration and production companies, typically in return for a fee and a performance or work obligation, such as *acquisition* of *seismic* data or drilling a well.

Exploration licenses are initially of limited duration (about 5 years) after which there might be a requirement to return half or more of the licensed acreage to the state. If hydrocarbons are discovered, a separate production license or production sharing agreement (PSA) is usually drawn up before development can proceed.

The term farm has varied definitions and on the Web one can find the following:

1. Where a company joins a joint venture in return for paying disproportionately for future joint venture operations—www.premier-oil.com/render.aspx
2. An arrangement whereby one working interest owner acquires an interest in a lease owned by another. Consideration for the transfer is usually an agreement by the transferee to pay all or part of the drilling and development costs, and the transferor frequently retains some interest—www.irs.gov/businesses/article/0,,id =201384,00.html

3. An arrangement whereby one oil operator "buys in" or acquires an interest in a lease or concession owned by another operator on which oil or gas has been discovered or is being produced—www.fusionenergy1.com/index.php/definitions/

The invitation to bid will come from regulatory agencies, which will offer a bid package—usually with some accompanying data (seismic and well data). Some agencies mount road shows, in which they will make presentations about the criteria for the bid, lease area, and blocks being offered, fiscal regime, signature bonuses, and minimal work program required as well as the percentage of local content in terms of services or investments. The process will have a deadline, after which the content of submitted bids will be revealed by the government in a public session.

The highest offer will result in the award of the block and the rights to explore.

10.2 FISCAL REGIMES—PSA AND TAX/ROYALTIES

There are over 100 fiscal systems in place around the world, some 50% of these being PSA, and 40% tax and royalties.

PSA regimes can be understood as an agreement between the parties to a well and a host country regarding the percentage of *production* each party will receive after the participating parties have recovered a specified amount of costs and expenses.

The company receives full recovery of production costs and a share of the remaining oil profits. Tax and royalties, on the other hand, mean that the company will pay a royalty on a fraction of the gross production and tax on net profits.

In some regions, such as for instance the GOM, some blocks may be subject to a royalty relief; in other words, the company will not have to pay a royalty until a certain volume has been produced (usually 87.5 MMboe). Subsequent to the attainment of this threshold volume, a royalty will be paid. This setup will make these blocks more attractive initially. In the GOM, the royalty is usually 12.5% or 18.75%. In other regions, for example in Brazil, royalties will vary from 10% up to 30% (new frontiers or the so-called presalt) (Figures 10.1 and 10.2).

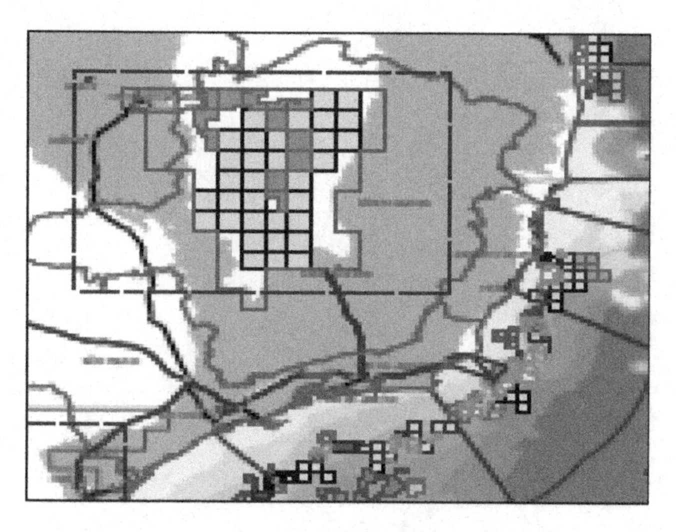

Figure 10.1 Example of blocks in a licensing round, South East Brazil.

GOM Lease Sale	Date of GOM Lease Sale	Water Depths (Meters)	Royalty Relief Bid Code	Fixed Royalty Rate	Royalty Suspension Volume (MMBOE of BCF)
CGOM SALE 213	3/17/2010	**400 to < 800**	RS16	18.75	**5 MMBOE**
CGOM SALE 213	3/17/2010	**800 to < 1,600**	RS17	18.75	**5 MMBOE**
CGOM SALE 213	3/17/2010	**1,600 to 2,000**	RS18	18.75	**12 MMBOE**
CGOM SALE 213	3/17/2010	**> 2,000**	RS19	18.75	**16 MMBOE**
CGOM SALE 208	3/18/2009	**400 to < 800**	RS16	18.75	**5 MMBOE**
CGOM SALE 208	3/18/2009	**800 to < 1,600**	RS17	18.75	**9 MMBOE**
CGOM SALE 208	3/18/2009	**1,600 to 2,000**	RS18	18.75	**12 MMBOE**
CGOM SALE 208	3/18/2009	**> 2,000**	RS19	18.75	**16 MMBOE**
CGOM SALE 208	3/18/2009	**0 to < 200**	RS14	18.75	**15 BCF**
CGOM SALE 208	3/18/2009	**0 to < 200**	RS14	18.75	**25 BCF**
CGOM SALE 208	3/18/2009	**0 to < 200**	RS14	18.75	**35 BCF**
CGOM SALE 208	3/18/2009	**200 to < 400**	RS15	18.75	**15 BCF**
CGOM SALE 208	3/18/2009	**200 to < 400**	RS15	18.75	**25 BCF**
CGOM SALE 208	3/18/2009	**200 to < 400**	RS15	18.75	**35 BCF**
WGOM SALE 207	8/20/2008	**400 to < 800**	RS16	18.75	**5 MMBOE**
WGOM SALE 207	8/20/2008	**800 to < 1,600**	RS17	18.75	**9 MMBOE**
WGOM SALE 207	8/20/2008	**1,600 to 2,000**	RS18	18.75	**12 MMBOE**
WGOM SALE 207	8/20/2008	**> 2,000**	RS19	18.75	**16 MMBOE**
WGOM SALE 207	8/20/2008	**0 to < 200**	RS14	18.75	**15 BCF**
WGOM SALE 207	8/20/2008	**0 to < 200**	RS14	18.75	**25 BCF**
WGOM SALE 207	8/20/2008	**0 to < 200**	RS14	18.75	**35 BCF**
WGOM SALE 207	8/20/2008	**200 to < 400**	RS15	18.75	**15 BCF**
WGOM SALE 207	8/20/2008	**200 to < 400**	RS15	18.75	**25 BCF**
WGOM SALE 207	8/20/2008	**200 to < 400**	RS15	18.75	**35 BCF**
CGOM SALE 206	3/19/2008	**400 to < 800**	RS16	18.75	**5 MMBOE**
CGOM SALE 206	3/19/2008	**800 to < 1,600**	RS17	18.75	**9 MMBOE**
CGOM SALE 206	3/19/2008	**1,600 to 2,000**	RS18	18.75	**12 MMBOE**
CGOM SALE 206	3/19/2008	**> 2,000**	RS19	18.75	**16 MMBOE**
CGOM SALE 206	3/19/2008	**0 to < 200**	RS14	18.75	**15 BCF**
CGOM SALE 206	3/19/2008	**0 to < 200**	RS14	18.75	**25 BCF**
CGOM SALE 206	3/19/2008	**0 to < 200**	RS14	18.75	**35 BCF**
CGOM SALE 206	3/19/2008	**200 to < 400**	RS15	18.75	**15 BCF**
CGOM SALE 206	3/19/2008	**200 to < 400**	RS15	18.75	**25 BCF**
CGOM SALE 206	3/19/2008	**200 to < 400**	RS15	18.75	**35 BCF**

Figure 10.2 Example of royalties and royalties relief in a lease sale in the GOM.

CHAPTER 11

The Value and Gains of Technology

The large accumulations of oil found in the past are not likely to occur again. However, several small size accumulations may still be awaiting discovery and a new large-scale play might still therefore be in the offing.

With global demand for oil rapidly increasing and easy oil becoming scarcer, we need to start looking at more difficult-to-exploit reservoirs from which to produce hydrocarbons.

To boost daily global oil supplies to 100 million barrels, against current levels of some 85 million barrels, will be "extremely difficult to reach" as the Total CEO said recently. It is not a question of available oil reserves, but a combination of technology, geopolitics, and actual production decline in existing fields. Geopolitics and depletion rates are highly complex variables upon which we do not have full control. However, technology can still be designed fit for purpose.

11.1 WHAT IS ENABLING TECHNOLOGY?

11.1.1 Makes Technology Able to Do Something for Us Now

Shallow reservoirs with biodegraded oil and better rock properties will be easier to produce than deeper ones with tight rock and light oil. Is technology going to help us here? Yes. World class reservoirs with high permeability and thickness with huge viscosities posed a challenge in past but today, we can design high HP submersible pumps to deal with heavy oil or inject polymers to reduce its viscosity. This is a reality already. Advanced research is in place to create special solvents that will increase ultimate recovery.

What about very deep targets? Down where the oil is lighter and the rock is much tighter it will be much more difficult to find a solution. Here is where we need innovation.

In ultra-deepwater and hostile regions such as the Arctic and the desert, as well as in new frontiers at depths over 4 km below the seafloor, new technologies will have to be created fit for purpose under thick salt walls. Technology will not only push these boundaries further but also rejuvenate mature assets.

11.1.2 The Roadmap to Technology

Figure 11.1 shows an overview of the current technologies in use or being pursued by the oil industry, including existing ones and emerging ones.

The driving force for integrated or independent oil and gas companies (IOC) is evaluating and implementing some new technologies for finding and producing oil and gas bearing in mind that these techniques need to be able to reduce costs, and increase the production and growth of the portfolio by extending the life of the reserves.

Here are some examples:

1. Drilling fast: For doing this we need to predict better the pressure regimes, shallow hazards, and rock types that we will be drilling. New measuring while drilling (MWD) techniques combined with forward-looking tools have been developed with capabilities to sample fluids at the same time as measuring data.
2. Unconventional deposits such as oil sands (at surface) + oil shale and tight gas (at subsurface) can hold billions of barrels of oils and trillions of cubic meters of gas. Again existing technology such as heating the rock or microfracturing the tight rock are in place onshore.

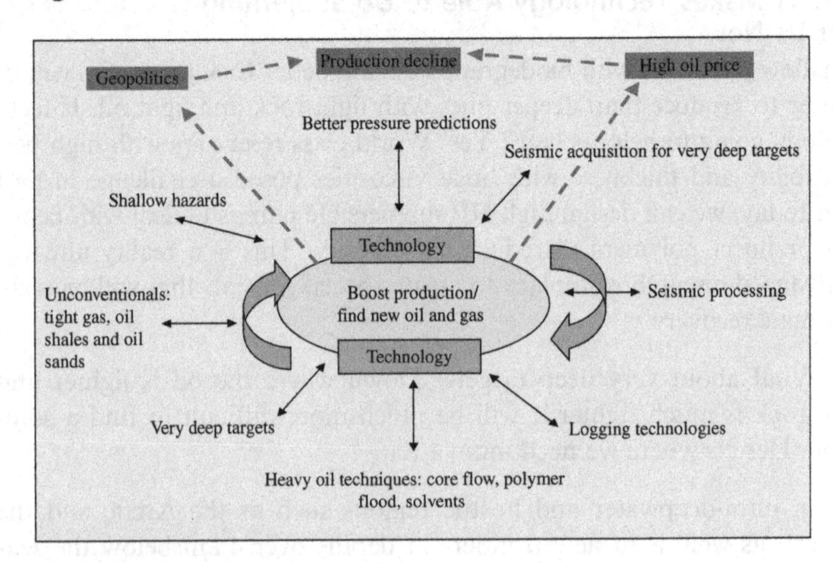

Figure 11.1 Roadmap to enabling technologies.

3. Unconventional geologic setting at deep-lying targets with low permeability: How to reach these horizons faster and how to achieve higher rates with the use of multistrings wells or long horizontal wells.
4. Electromagnetic methods (EM) and improved seismic acquisition and processing methods to achieve sharper images and validate and differentiate traps and fluids.
5. Revitalized old concepts such as deployment of surface blow out preventers (BOPs) to reduce weight on risers or monodiameter wells.

11.1.3 Technology + Innovation

Innovation is crucial to start a new technology together with creativity. Figure 11.2 shows the evolution of innovations by offshore companies in terms of deepwater systemology. As one can see, the energy sector is investing in innovation and it has proven to pay off.

Again IOC have been saving millions of dollars yearly by decreasing their exploration and development costs through the use of innovation technologies. To cite two simple examples:

1. Giving old data new life by taking data originally acquired in the 1960s and stored in hard copy or tape and converting them to digital format that can be used in modern interpretations tools.

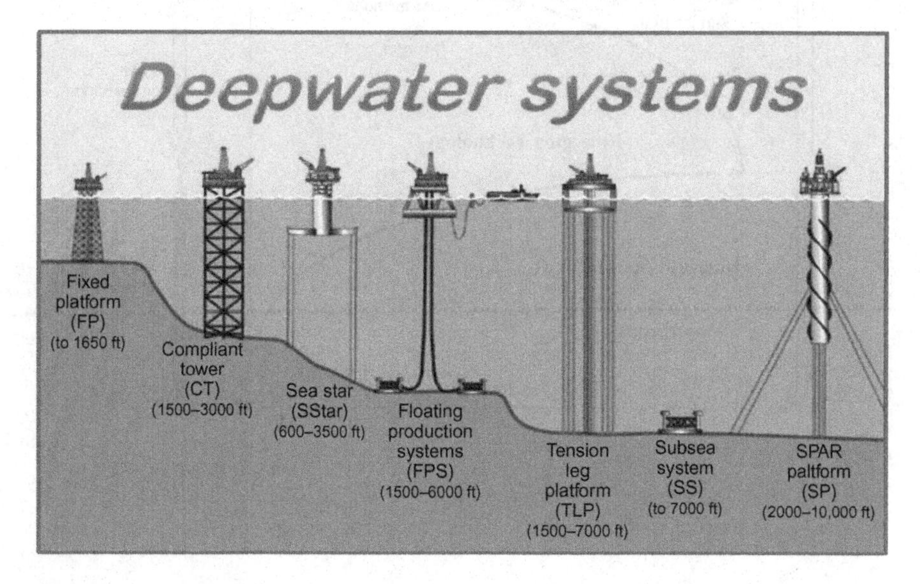

Figure 11.2 Evolution of offshore structures.

2. The development of smart wells and digital fields that can be operated remotely.

11.1.4 Emerging Technologies

Figure 11.3 shows interesting aspects of hot spots for technology innovations. In offshore areas, we can see that the level of technology required will depend on the nature of the major drivers for achieving production, such as fluid or rock properties. Shallow depths below the mud line will be more challenging in terms of biodegraded fluids (more viscous oil) than deep targets where the rock properties may be blocking economic production of light fluids.

Figure 11.3 shows exactly this. Existing technologies for drilling, completion, and production are able to deliver together with some emerging technologies up to a certain depth. Beyond that only innovative technology will be able to do the job. Of course, we do not have yet a technology to produce economically very deep targets due to lack of producibility for the reservoirs or expensive drilling.

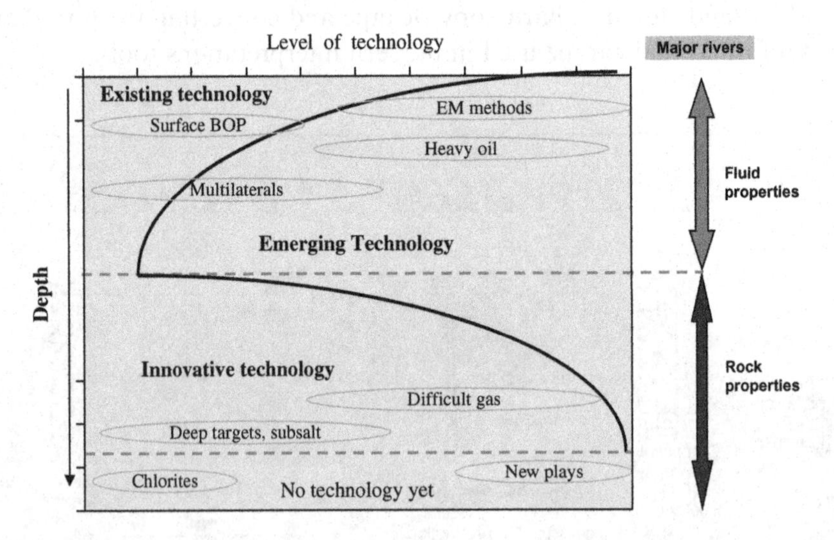

Figure 11.3 Technology chart.

11.1.5 What Should We Expect?

We should expect in the near future more innovation and creativity.

1. Outsourcing from other industries such as biochemicals, light materials, and nanotechnology.
2. Simple techniques such as use of the surface BOP as illustrated in Figure 11.4.
3. Slim/monodiameter wells with no logging, or evaluate tools where the only objective is to get a sample of oil/gas shows.
4. More cluster developments where small fields will undergo tieback together in order to make their development feasible.
5. Fast road from discovery to appraisal, development, and production perhaps using discovery wells as appraisals and producers.
6. Wells with smart completions.
7. Basin modeling work and geochemistry, in order to predict migration path, accumulations, and fetching areas as well as fluid properties.

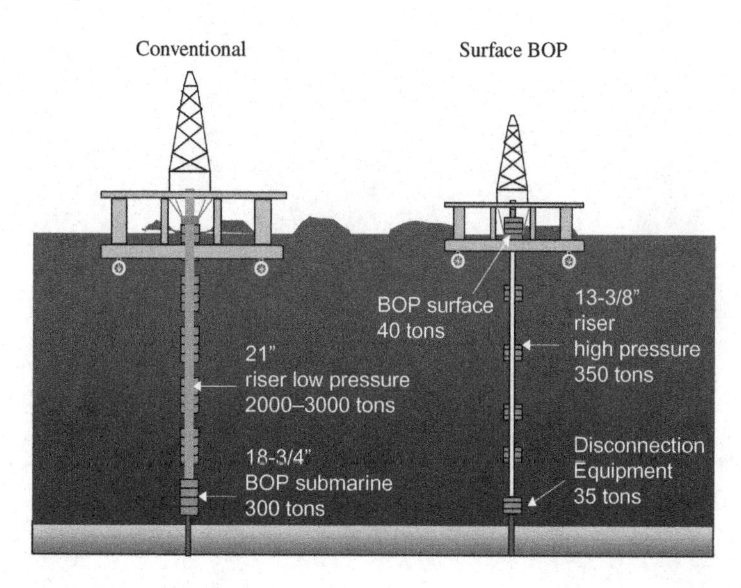

Figure 11.4 Surface and sea bottom BOPs.

CHAPTER 12

Field Case Evaluations

This chapter will show 15 deepwater field cases on economic analysis covering different types of oil and gas reservoirs at several depths in the *Gulf of Mexico, one case for Brazil, one case for Southern North Sea* as well as distinct field development options such as standalone hosts and tie-backs. The 15 cases will be the following:

1. Small accumulation with heavy oil
2. Big accumulation with heavy oil
3. Small accumulation with light oil
4. Big accumulation with light oil
5. Small accumulation with dry gas
6. Big accumulation with dry gas
7. Small accumulation with gas and condensate
8. Big accumulation with gas and condensate
9. Accumulation with oil and gas reservoirs combined
10. Example of fields evacuated to a shared facility
11. Example of a tight gas offshore reservoir
12. Example of a very large offshore oil field
13. Example of an offshore field with multilayer reservoirs
14. Example of field optimization with horizontal wells
15. Field unitization example

12.1 CASE 1—SMALL ACCUMULATION <500 MMbbl IN PLACE WITH HEAVY OIL

Make an economical analysis for an oil prospect with a volume in place of 500 MMbbl of oil, low dissolved gas, situated in deepwater Gulf of Mexico with a water depth of 5500 ft, Paleogene age and total depth of 25,000 ft with an area of 3420 acres. The net pay is approximately 300 ft. No exploratory wells, and all data is provided from seismic.

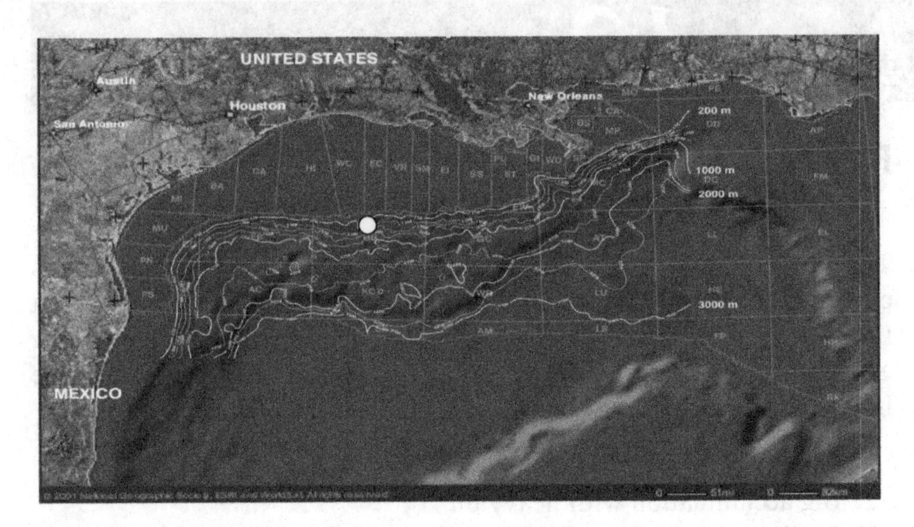

Figure 12.1 Field location map.

12.1.1 Recovery Factor Estimation

The first thing will be to estimate the recover factor for this accumulation, In the example shown in the Recovery factor chapter it is possible to see a trend of recovery factor by depth and by play. We can see that we should expect recoveries of 25% for a Paleogene play with reservoirs at depth of 25,000 ft. Therefore, our recoverable volume will be 25% × 500 equaling 125 MMbbl.

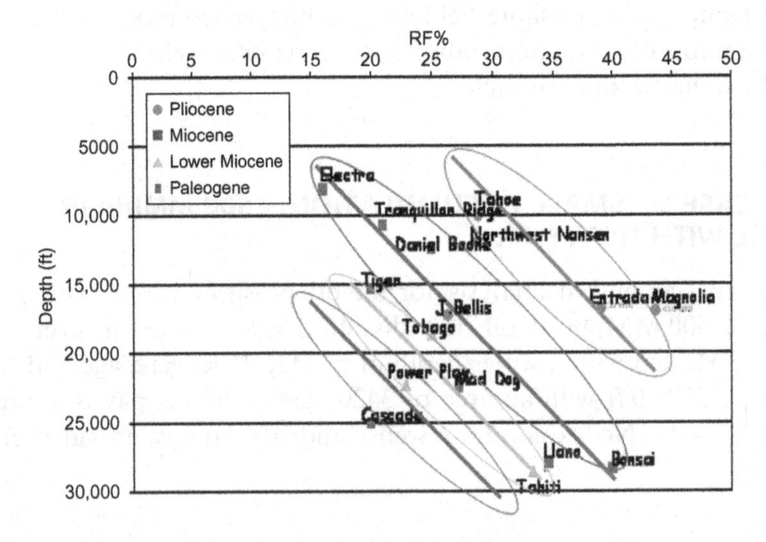

Figure 12.2 Example of a trend deriving using analogs and grouping parameters.

12.1.2 Rock/Fluid Properties and Well Count Estimation

Next step is to derive the number of wells and their initial rate. To do this we need first to estimate some rock and fluid properties.

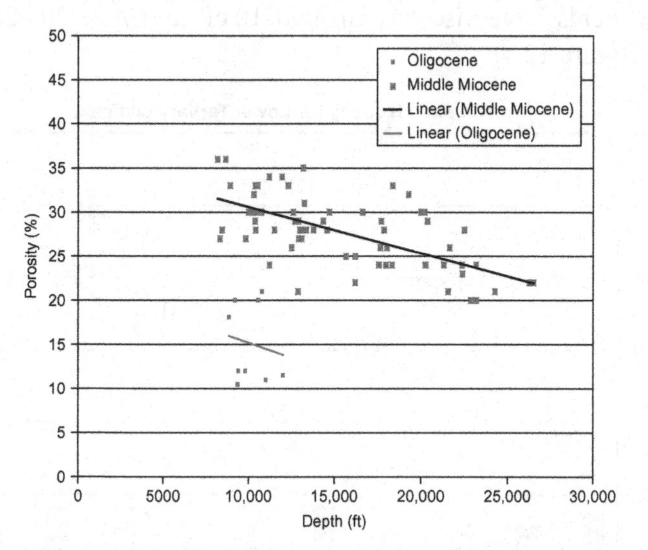

Figure 12.3 Trend for porosity vs. depth.

First, let's get an estimate for porosity. Looking at the dots for Oligocene (Upper Paleogene), we can see that for a depth of 25,000 ft, there is no data. However, we see that at shallower depths the porosity varies between 10–20%. Using this range we can estimate permeability such as 5–75 mD.

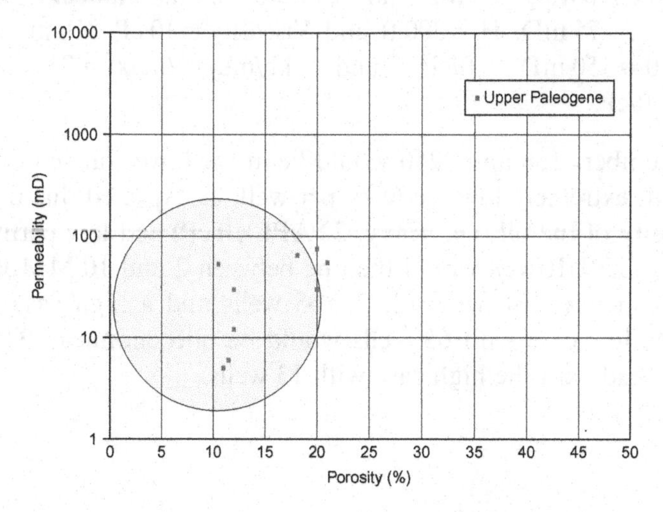

Figure 12.4 Permeability for Oligocene play from a database with analog information.

We have estimated so far recovery factor, porosity range and permeability ranges using analog information. It is missing some fluid information such as viscosity and API grade for the oil. Most of the Paleogene fields have viscosity around 10 cP and API 20–22 as suggested in Figure 12.5.

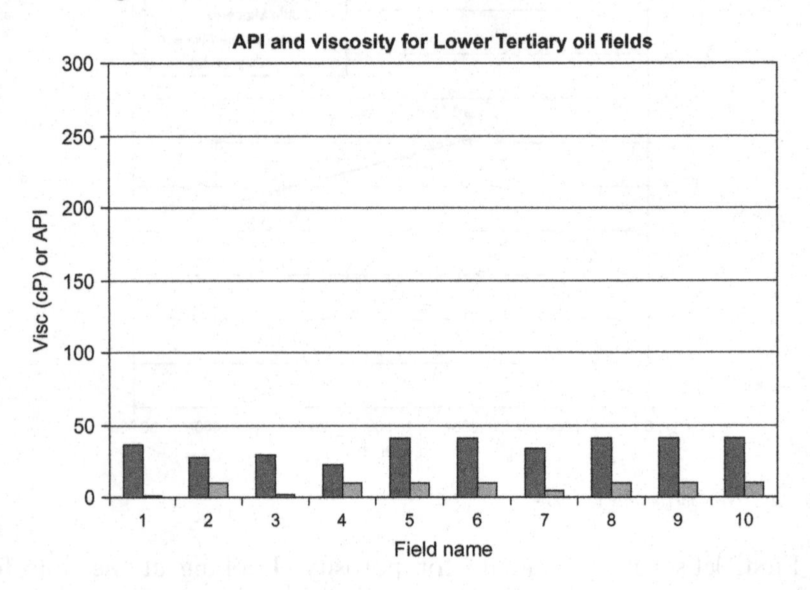

Figure 12.5 Viscosity for fields in the Lower Tertiary.

With K, H and Viscosity, we can now calculate the index kh/mu and estimate the recovery per well and therefore the total number of producer wells. K = 5–75 mD, H = 300 ft and Viscosity = 10 cP. Kh/mu (min) = 5 × 300/10 = 150 mD ft/cP and kh/mu (max) = 75 × 300/10 = 2250 mD ft/cP.

The numbers 150 and 2250 would lie in the lower range (see Figure 12.5), with extremely low recovery per well, as expected due mainly to the viscosity of the oil (i.e. heavy, 22 API, 10 cP) and low permeability reservoir. The UR/well would then be between 2 and 10 MMbbl of oil leading to a low case with 125/2 = 63 wells and a high case with 13 wells. The low case with 63 wells would be uneconomical. Therefore, let's go ahead with the high case with 13 wells.

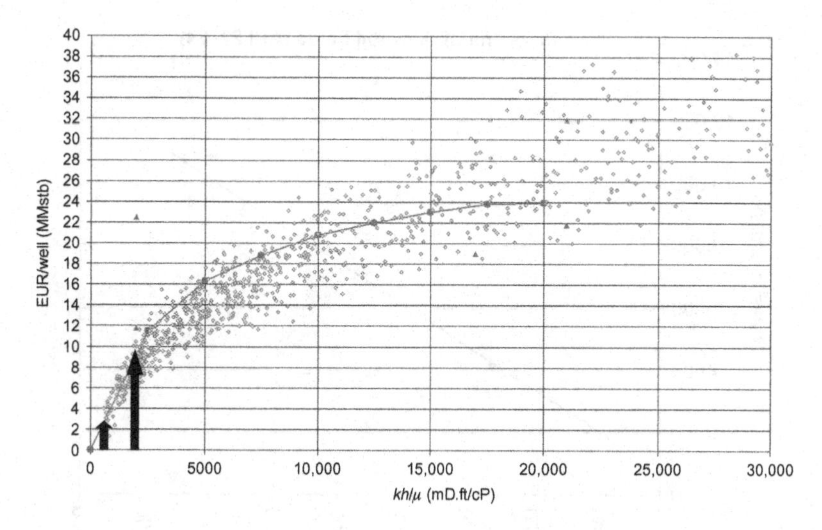

Figure 12.6 Estimation of number of wells based on rock and fluid properties.

So far, we have estimated the following parameters for this project: Number of Wells—13, well spacing = 3420 acres/13 = 263 acres, Permeability—5–75 mD, Viscosity = 10 cP, Recovery per well = 2–10 MMbbl, H = 300 ft, KH = 1500–22500 mD × ft.

To estimate our facility type and capacity we need to calculate the peak production rates for each of the 13 wells and generate a production curve with time.

12.1.3 Well Initial Rates and Notional Production Profile

The best way to estimate well initials is using the properties we have already found—K (permeability), H (thickness) and Fluid Viscosity.

Figure 12.7 is derived from analogs (oil fields in the GOM with API averaging around 24 API). Entering with KH equaling 22,500, Qi is about 2200 bbl/d. Note that lighter oils would produce higher rates.

So 13 wells at 2200 bbl/d would yield 28,600 bbl/d. We now need to create a production curve that produces a plateau of 28,600 bbl/d and a total reserves of 125 MMbbl.

Figure 12.8 shown on the following page would fit as our production scenario. In the first year, the production is lower because not all wells will be on production (in this case, only half of the total so about 6 wells).

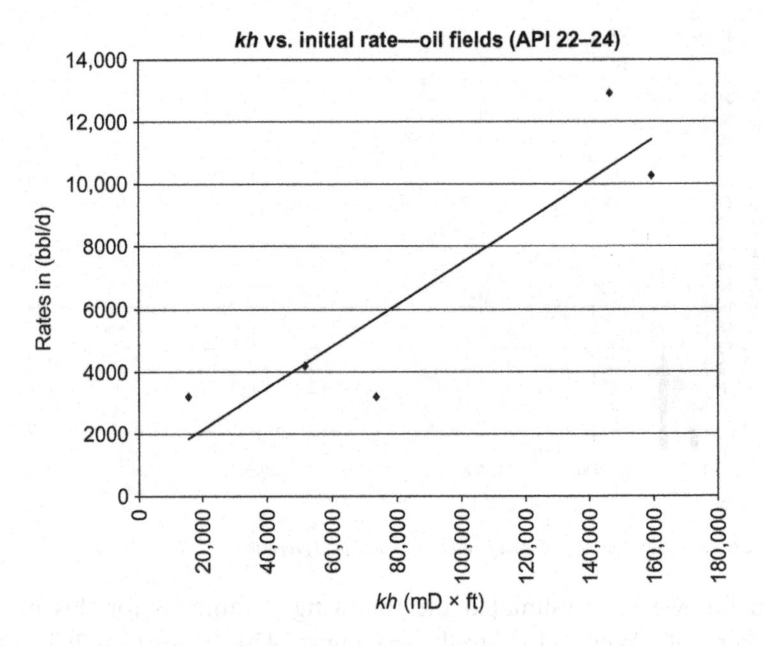

Figure 12.7 Estimation of well rates based on rock and fluid properties.

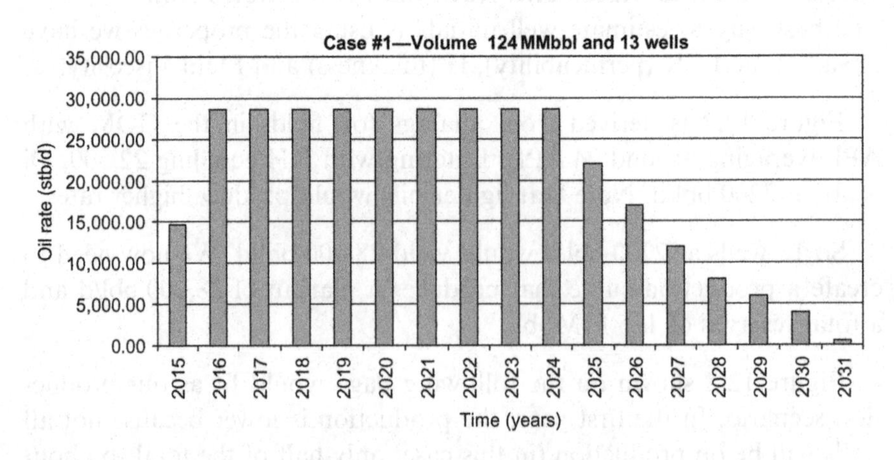

Figure 12.8 Production forecasting.

12.1.4 Costs: Drillex, Capex and Opex

Now it is necessary to estimate the cost to drill the wells, build the facilities, and evacuate the production as well as the operating costs.

12.1.5 Drilling Costs

Drilling costs will depend on the depth of the well and the daily rig rate. The rig daily rate will vary according to the rig type, water depth, distance from shore and drilling depth. For onshore, it will be <100,000 $/day, and for deepwater offshore Gulf of Mexico, it can be very high—up to 600,000 to 800,000 $/day (values are from 2010). The number of days will be a function of depth. For usual depth up to 20,000 ft, we can assume 70 to 80 days and for deeper depths up to 32,000 ft, a maximum of 150 days.

We can then create the following (Table 12.1) to estimate the dry hole cost:

Table 12.1 Drilling Costs as a Function of Rig Rate and Drilling Time

2010 Rig Rate (MMS/day)	Depth (ft)	Drilling Days	Well Cost (MM$)
500–800	20,000	70–80	35–64
500–800	26,000	110	55–88
500–800	32,000	150	75–120

This is the cost of the exploratory well. For the appraisal and developments wells we will need to add some extra costs such as evaluation and completion costs. Our volume is 125 MMbbl of oil (so will need one exploratory well plus 1 appraisal well and 13 development wells).

One exploratory well will cost 55–88 MM$, and 1 appraisal well will cost 55–88 MM$ plus evaluation costs of 15 days more totaling 62.5–100 MM$.

One development well will cost 55–88 MM$ plus completion costs (+80%) totaling 99–158.4 MM$. Remember that we will need 13 wells. We may assume that we learn costs while we drill more wells, so let's assume the first 2 wells will cost the 99–158.4 MM$ and the remaining 11 wells will be slightly cheaper.

12.1.6 Facilities and Subsea Costs (Capex)

Figure 12.9 shows that topsides and hull for semis (semi-submersible production units) may be seen as a function of its total capacity and can be considered approximately linear. The numbers were made up to serve as an example and may not be accurate enough.

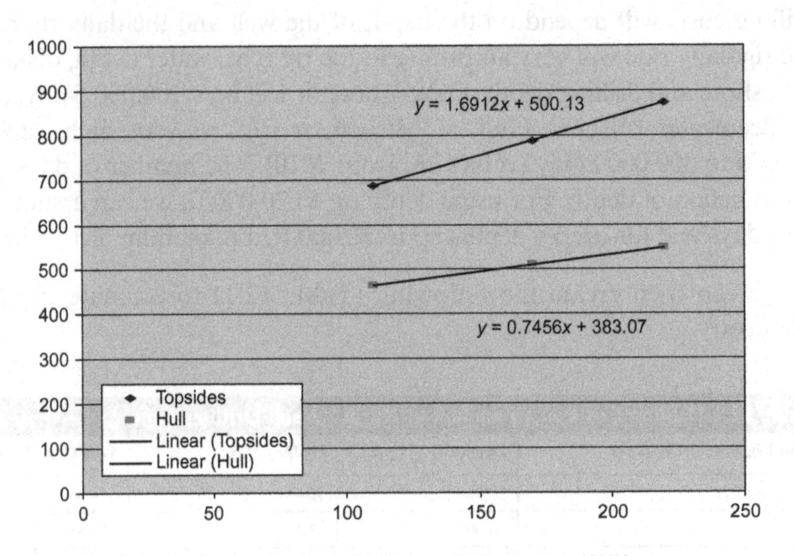

Figure 12.9 Hull and topsides Capex estimation in MM$ versus FPU capacity in kbbl/day.

The x-axis represents unit capacity in thousands of barrels per day and the y-axis is the cost (i.e. fabrication plus installation) in MM$. We want our unit to be sized to accommodate 30 kbbl/d. Using the equations as shown in Figure 12.9 we get for 30 kbbl/d, hull cost is about 400 MM$ and top side is about 552 MM$, with a total of 958 MM$. We may want to add some contingency costs here around 25% or 1278 MM$.

Next step is to calculate the subsea costs and export pipelines costs. For 13 wells, we will need two manifolds (one with 6 wells and another one for 7 wells) plus inter field flow lines and risers. These costs come from usually from a previous project and/or vendors. Let's assume some factors here to estimate these costs:

- Well templates and manifolds = 34 MM$/well
- SCR Production Riser = 2 risers per manifold = 4 Risers at 0.35 MM$/(″ × 1000 ft)

- SCR Export Riser = 1 at 0.28 MM$/(″ × 1000 ft)
- Flow lines and Export Pipelines at 0.23 and 0.16 MM$/ (″ × miles). Let's assume flow lines length equals 5 miles and export pipeline length equals 30 miles

These are estimated factors as provided by vendors and publications. Templates and manifolds are 34 × 13 wells equaling 442 MM$, and production risers are 4 × 8 × water depth (i.e. 1.5 × 5500 ft) × 0.35/1000 × 10″ equaling 924 MM$. Export riser is 0.28 × 5500 ft/1000 ft × 10″ equaling 15.4 MM$, and FL plus export pipelines is 0.23 × 5 miles + 0.16 × 30 miles × 8″ equaling 48 MM$.

12.1.7 Operating Costs (Opex)

The operating costs are more complex and difficult to be estimated using only charts or factors. It is usually a function of the peak production rate and tariff rates paid to export production via third party pipelines.

Table 12.2 is tentative to summarize Opex numbers per BOE (barrels of oil equivalent) for the years on plateau or peak rates based on what has been published by operators in the GOM.

Table 12.2 Total OPEX Estimation

Facility Type	No tariff	With Tariff
Standalone FPU	8–15 $/boe	12–25 $/boe
Tie backs	5–10 $/boe	8–13 $/boe

So for this project (standalone with Tariff), we will estimate our Total Opex for the field life as a minimum of 125 MMbbl × 12 and maximum of 125 MMbbl × 25 equaling 1500 MM$ and 3125 MM$ for the 9 years of production on plateau. This total cost should be prorated proportionally to production for the remaining years.

For the peak rates on plateau, our yearly Opex would be 1500/9 or 3125/9 and prorated for the remaining years. A minimum of 25% Opex is recommended for the years where the production is lower than 25% of peak to take into account the maintenance costs and other costs that are not a direct function of the production.

12.1.8 Project Timing and Input to Economic Analysis

We have now calculated most of the input we need for our economic analysis and to the end of our VV (volume to value) process (Table 12.3).

Table 12.3 Timing Period from Exploratory Well to First Production

Facility Type	Time in Years
Standalone FPU	1 year—Exp well, 2 year—App well 3–7 years (facilities + subsea)
Tie backs	As above Up to 3 years for the facility tie back

The timing of the project is crucial as all the other inputs but it is also flexible depending how aggressive the company wants to drill the wells and invest in advance on the fabrication of facilities.

Ideally, we will drill at least one well per year and consider a period of time to fabricate the facilities and have it in place.

The table provides an overview for the total project time from exploration to first production. In the best case, this time won't be less than 5 years. For a big project, it can reach 10 or more years from the exploratory well to the first production.

Now it is possible to place all the information we need to calculate net present value NPV for the opportunity and other economic parameters. Figure 12.10 summarizes the inputs that any economic package will require, basically the revenues (volumes) and the expenditures. Note that in the Figure 12.10, there are min and max numbers for the costs. The user may choose to pick an average number before running the economics.

Date	Oil Production rate	Gas Production rate	Exploration Drilling	Appraisal Drilling	Development Drilling	Capex	Capex	Capex	Abandonment	Opex	Comments
	stb/day	MMscf/day	MM$	MM$	MM$	MM$	MM$	MM$	MM$	MM$	
			min-max	min-max	min-max	Facility	Subsea	pipeline			
2010			55-88								1 Exploratory
2011				62.5-100	(99-158) × 1	255.6					1 Appraisal+1 well
2012					(99-158) × 3	255.6					3 wells
2013					(99-158) × 3	255.6	147.3333	924			3 wells
2014					(99-158) × 3	255.6	147.3333	15.4			3 wells
2015	14,300.00				(99-158) × 3	255.6	147.3333	48			3 wells
2016	28,600.00									83.33-173.61	
2017	28,600.00									166.66-347.22	
2018	28,600.00									166.66-347.22	
2019	28,600.00									166.66-347.22	
2020	28,600.00									166.66-347.22	
2021	28,600.00									166.66-347.22	
2022	28,600.00									166.66-347.22	
2023	28,600.00									166.66-347.22	
2024	28,600.00									166.66-347.22	
2025	22,600.00									83.33-173.61	
2026	17,000.00									41.67-86.81	
2027	12,000.00									41.67-86.81	
2028	8,000.00									41.67-86.81	
2029	6,000.00									41.67-86.81	
2030	4,000.00									41.67-86.81	
2031	1,000.00								127.60	41.67-86.81	10% of Total Facility Capex
Total	volume = 125 MMbbl		55-88	62.5-100	1287-2054	1276	442				

Figure 12.10 Economic analysis inputs.

12.2 CASE 2—BIG ACCUMULATION >500 MMbbl IN PLACE WITH HEAVY OIL

Make an economical analysis for a oil prospect, with a volume in place of 1000 MMbbl of oil, low dissolved gas, situated in a remote region in deepwater Gulf of Mexico with a water depth of 5500 ft.

It is of Lower Tertiary (Paleogene) age and situated at a total depth of 25,000 ft with an area of 6800 acres. Net pay will be approximately 300 ft. No exploratory wells, and all data is provided from seismic.

For this case all the assumptions calculated in the case study 1 will apply in terms of RF, EUR per well, well initials, rock and fluid properties. The number of wells will increase now to 25 wells (UR/well is about 10 MMbbl) because we have a higher volume now.

The production profile will be longer too. We need now to create a production curve that produces a plateau of about 60 kbbl/d and yield cumulative production of 250 MMbbl. Figure 12.11 would fit as our production scenario. In the first year, the production is lower because not all wells will be on production, in this case only half of the total or about 12 wells.

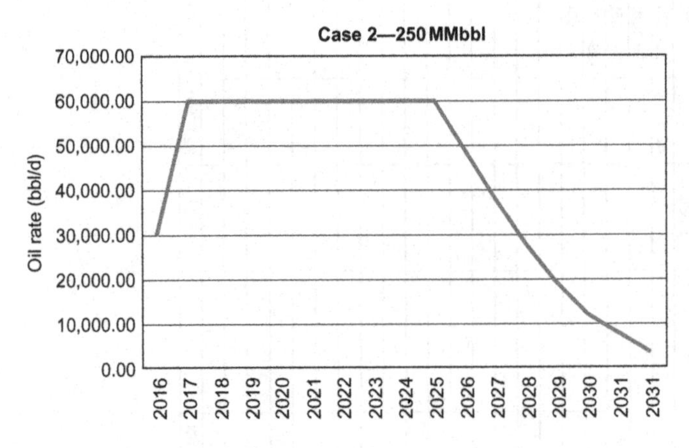

Figure 12.11 Production forecasting.

12.2.1 Costs: Drillex

The cost to drill the wells is the same as in case study 1. However, now we will have more appraisals (and development) wells due to higher volumes. Now we may require two drilling rigs working at the same time as the number of well has increased. Also the facilities will double as well as the operating costs. We will need at least two appraisal wells and 25 development wells.

12.2.2 Facilities and Subsea Costs (Capex)

Figure 12.12 will be used again to provide an estimation of Capex.

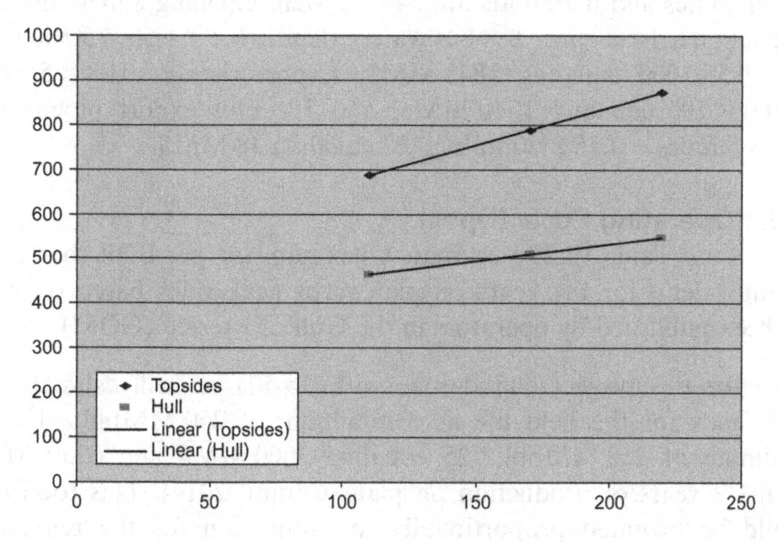

Figure 12.12 Hull and topsides Capex estimation MM$ versus FPU capacity.

We want our unit to be sized to accommodate 60 kbbl/d. Using the equations as shown in Figure 12.12 we get for 60 kbbl/d, hull cost is about 450 MM$ and top side is about 650 MM$ totaling 1100 MM$. We may want to add some contingency costs here around 25% or 1375 MM$.

Note that although the rates doubled the Capex costs did not double.

Next step is to calculate the subsea costs and export pipelines costs. For 25 wells, we will need four manifolds (three with 6 wells and another one for 7 wells) plus inter field flow lines and risers. Let's assume that some of the same factors used in case study 1 to estimate these costs:

- Well templates and manifolds = 34 MM$/well
- SCR Production Riser = 2 risers per manifold = 8 Risers at 0.35 MM$/(″ × 1000 ft)

- SCR Export Riser = 1 at 0.28 MM$/(″ × 1000 ft)—assumed that the riser has enough capacity to handle the increase on production.
- Flow lines and Export Pipelines at 0.23 and 0.16 MM$/ (″ × miles). Let's assume flow lines length = 5 miles and Export Pipeline length = 30 miles (let's keep to the same as for example 1 assuming the flow lines can handle the increase of rate up to 60 kbbl/d).

Templates and manifolds are 34 × 25 wells equaling 850 MM$, and production risers are 8 × 8 × water depth (i.e. 1.50 × 5,500 ft) × 10″ × 0.35/1000 equaling 1848 MM$. Export riser is 0.28 × 5500 ft/ 1000 ft × 10″ equaling 1540 MM$, and FL plus export pipelines is 0.23 × 5 miles + 0.16 × 30 miles × 8″ equaling 48 MM$.

12.2.3 Operating Costs (Opex)

We will use Table 12.2 to estimate Opex numbers per BOE (barrels of oil equivalent) for the years on plateau or peak rates based on what has been published by operators in the Gulf of Mexico (GOM).

So for this project (standalone with Tariff) we will estimate our total Opex for the field life as a minimum of 250 MMbbl × 12 and maximum of 250 MMbbl × 25 equaling 3000 MM$ and 6250 MM$ for the 9 years of production on plateau (until 2014). This total cost should be prorated proportionally to production for the remaining years.

For the peak rates on plateau, our yearly Opex would be 3000/9 or 6250/9 and prorated for the remaining years. A minimum of 25% Opex is recommended for the years where the production is lower than 25% of peak to take into account the maintenance costs and other costs that are not a direct function of the production.

12.2.4 Project Timing and Input to Economic Analysis

We have now calculated most of the input we need for our economic analysis and to the end of our VV (volume to value) process. The timing of the project (first production) will be longer than one year than in case study 1 because we will need an extra appraisal well as the volume is now two times bigger. This second appraisal well may be drilled in the second or third year pushing all the developments well one year later and thus delaying the start up of first production. We will also assume that the extra development wells will be drilled simultaneously by a second rig.

Date	Oil Production rate	Gas Production rate	Exploration Drilling	Appraisal Drilling	Development Drilling	Capex	Capex	Capex	Abandonment	Opex	Comments
	Stb/day	MMScf/day	MM$	MMS	MM$	MM$	MM$	MM$	MM$	MM$	
			min-max	min-max	min-max	Facility	Subsea	pipeline			
2010			55-88								1 Exploratory
2011				62.5-100		275					2 Appraisal+1 well
2012				62.5-100	(99-158) × 1	275					6 wells (2 rigs)
2013					(99-158) × 6	275	283.33	1848			6 wells
2014					(99-158) × 6	275	288.33	15.5			6 wells
2015					(99-158) × 6	275	288.33	48			6 wells
2016	30,000.00				(99-158) × 6					166.50	
2017	60,000.00									333.00	Opex max with Tariff
2018	60,000.00									333.00	
2019	60,000.00									333.00	
2020	60,000.00									333.00	
2021	60,000.00									333.00	
2022	60,000.00									333.00	
2023	60,000.00									333.00	
2024	60,000.00									333.00	
2025	60,000.00									333.00	
2026	48,500.00									333.00	
2027	37,000.00									166.50	
2028	27,000.00									166.50	
2029	19,000.00									166.50	
2030	12,000.00									83.25	
2031	7,500.00									83.25	
2032	3,700.00								137.50	83.25	10% of Total Facility Capex
Total			55-88	125-200	2475-3950	1375	850	1911.5			

Figure 12.13 Economic analysis inputs.

12.3 CASE 3—SMALL ACCUMULATION <300 MMbbl IN PLACE WITH LIGHT OIL

Make an economical analysis for an oil prospect, with a volume in place of 300 MMbbl of oil, low dissolved gas, situated in deepwater Gulf of Mexico with a water depth of 2500 ft. It is Lower Miocene age and total depth of 25,000 ft with an area of 1200 acres. Net pay will equal 100 ft. No exploratory wells, and all data is provided from seismic. The prospect is just 15 miles of the existing Marco Polo Platform.

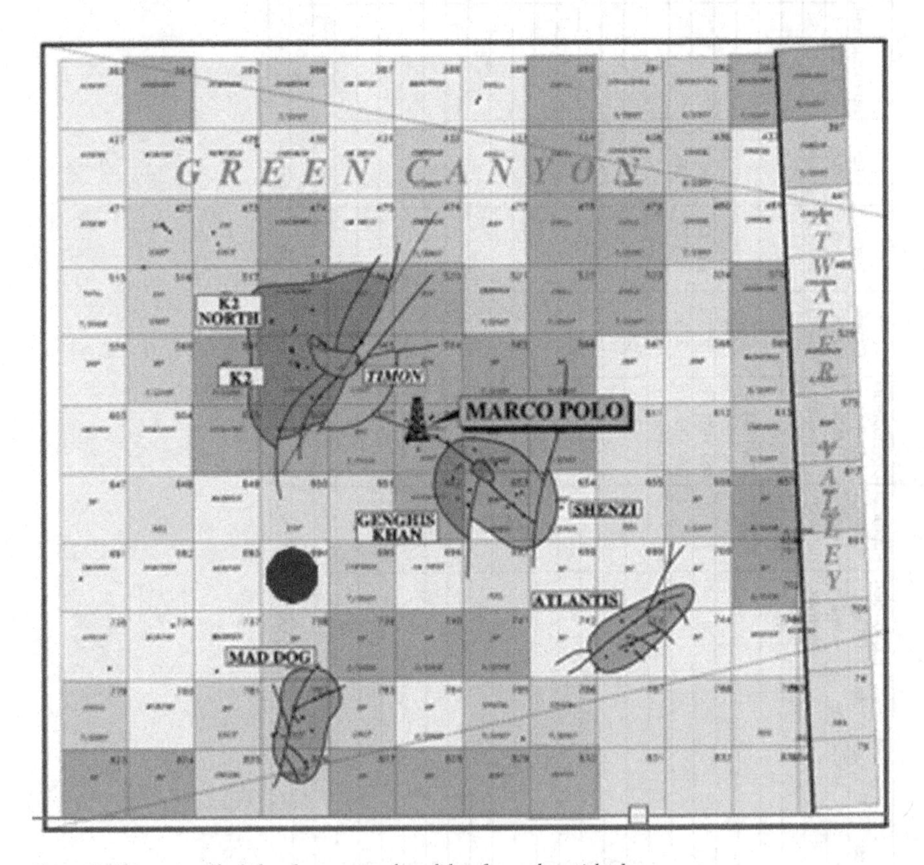

Figure 12.14 Location of a light oil prospect and candidate for a subsea tieback.

This example is a good case for a subsea tieback due to the small volume and also the proximity of an existing infrastructure.

12.3.1 Recovery Factor Estimation

The first thing again will be to estimate the recover factor for this light oil accumulation. We can see we should expect recoveries of 30% for a Miocene play with reservoirs at depth of 25,000 ft. Therefore, our recoverable volume will be 30% × 300 or 90 MMbbl.

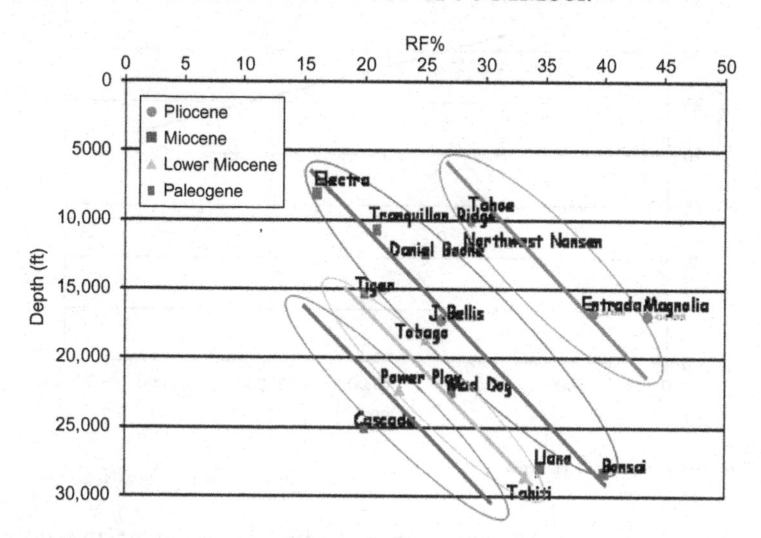

Figure 12.15 Example of a trend deriving using analogs and grouping parameters.

In this case, it would be prudent to look at ranges of expected recovery from the nearby fields such as Mad Dog and Genghis Khan. Commercial databases and searching on the literature suggests Mad Dog at 25–32% RF and Genghis Khan at 16.2%. Although our 30% estimation is not bad, we could also run a lower case for a 16% recovery factor 16 % × 300 equaling 48 MMbbl.

12.3.2 Rock/Fluid Properties and Well Count Estimation

Next step is to derive the number of wells and their initial rate. To do this, we need first to estimate some rock and fluid properties.

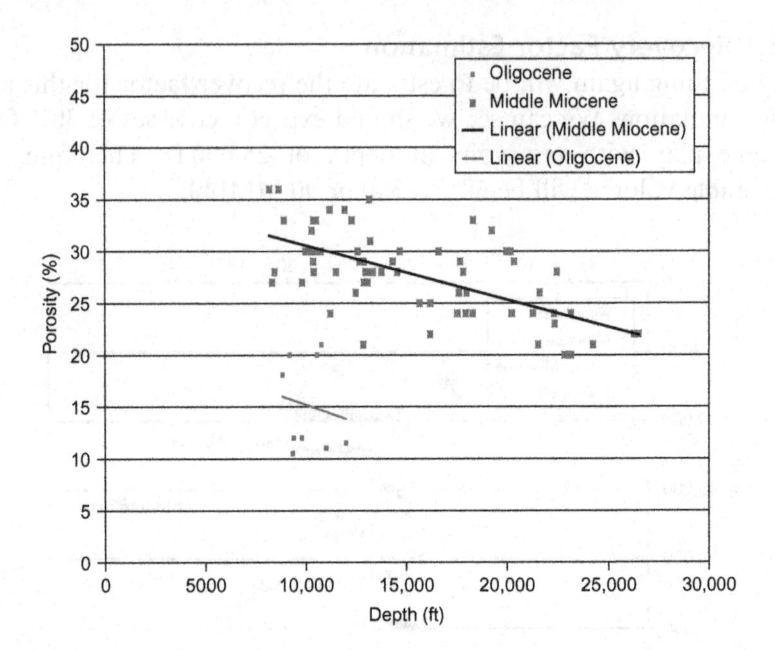

Figure 12.16 Trend for porosity vs. depth.

First, let's get an estimate for porosity. Looking at the dots for Miocene we can see that for a depth of 25,000 ft porosity is around 22% and permeability from 100 to 500 mD.

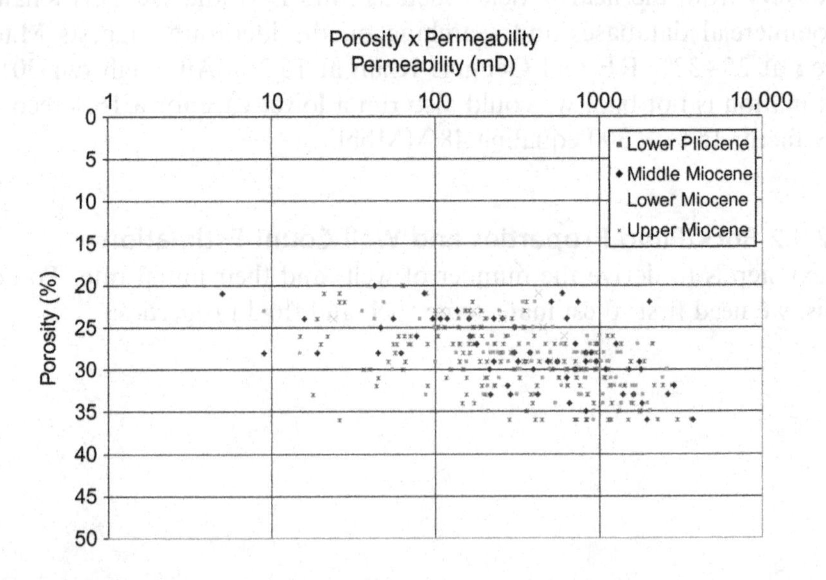

Figure 12.17 Porosity versus permeability trend.

We have estimated so far recovery factor, porosity range and permeability ranges using analog information. It is missing some fluid information such as viscosity and API grade for the oil.

Again let us use Mad Dog fluid as an analog. Mad Dog has API 32 and a viscosity of 2 cP.

With K, H and Viscosity we can calculate now the index kh/mu and estimate the recovery per well and therefore the total number of producer wells. K = 100–500 mD, H = 100 ft and Viscosity = 2 cP. Kh/mu (min) = 100 × 300/2 = 15,000 mD ft/cP and kh/mu (max) = 500 × 100/2 = 25,000 mD ft/cP.

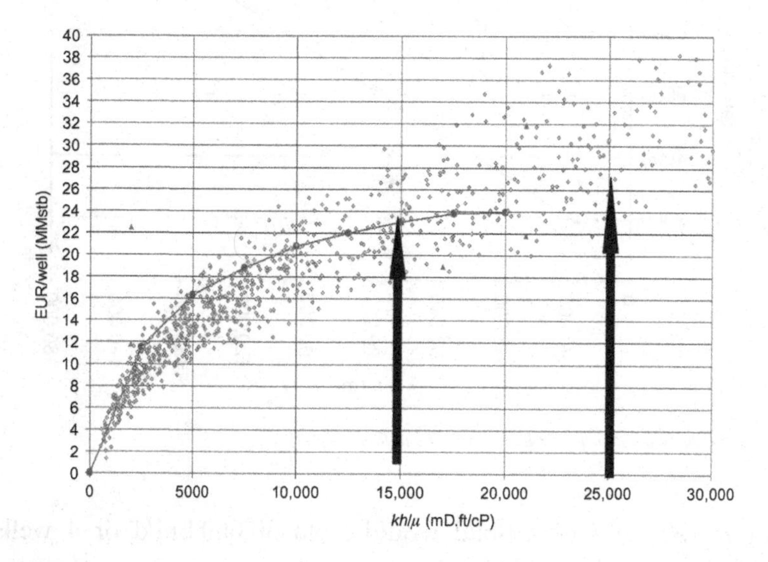

Figure 12.18 Estimation of number of wells based on rock and fluid properties.

The UR/well would then be between 23 and 28 MMbbl of oil (average 25 MMbbl) leading to a low case with the low volume 48/25 or 2 wells and a high case with 90/25 or 4 wells.

12.3.3 Well Initial Rates and Notional Production Profile

The best way to estimate well initials is using the properties we already found: K (permeability), H (thickness) and Fluid Viscosity. Figure 12.19 is derived from analogs (oil fields in the GOM with API averaging around 24 API). Our fluid will have a higher API (32) and lower viscosity because we are being conservative.

Entering KH equaling 15,000, Qi is about 10,000 bbl/d. For KH equaling a 25,000 rate, Qi could be up to 20,000 bbl/d. Let's use the average 15,000 bbl/d.

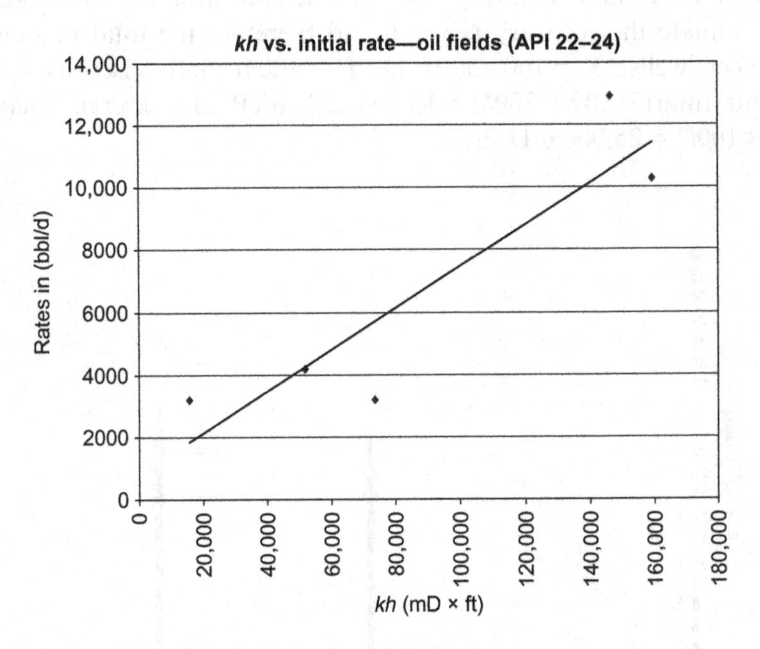

Figure 12.19 Estimation of well rates based on rock and fluid properties.

So 2 wells at 15,000 bbl/d would yield 30,000 bbl/d or 4 wells at 15,000 equals 60,000 bbl/d.

We now need to create a production curve that produces a plateau of 30,000 bbl/d or 60,000 bbl/d and total reserves of 90 MMbbl or 48 MMbbl for the low case. The one shown in Figure 12.20 would fit as our production scenario for the 60,000 bbl/d with 4 wells.

12.3.4 Costs: Drillex, Capex and Opex

Now it is necessary to estimate the cost to drill the 2 wells, and evacuate the production as well as the operating costs.

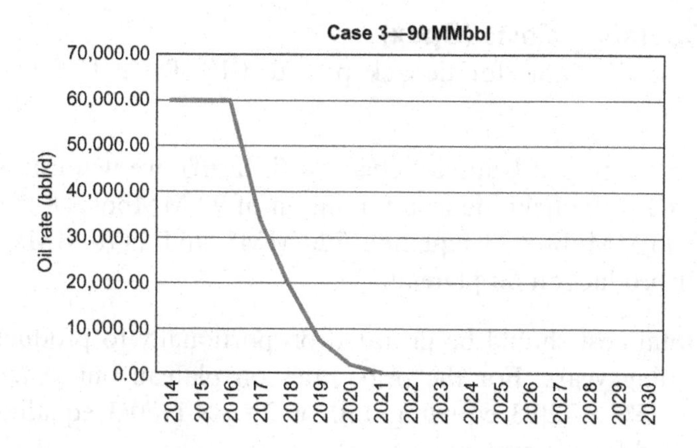

Figure 12.20 Production forecasting.

12.3.5 Drilling Costs

Using again Table 12.1 to estimate the dry hole cost.

For a depth of 25,000 ft the drilling costs will be 55–88 MM$ for the exploratory well. As the volumes are lower than 100 MMbbl, we will not need an appraisal well. One development well will cost 55–88 MM$ plus completion costs (+80%) totaling 99–158.4 MM$. Note we will need 2 wells (or 4 in the high case scenario).

12.3.6 Facilities and Subsea Costs (Capex)

In this example, we will have only subsea and pipeline costs as there is no hub or facility to be built because we will tie back production to Marco Polo Platform. Next step is to calculate the subsea costs and export pipelines costs. For 2 or 4 wells we will need only one manifold plus export pipeline to the Marco Polo Platform and the riser. Costs should look like:

- Well templates and manifolds = 34 MM$/well
- SCR Production Riser = at 0.35 MM$/(″ × 1000 ft)
- Export Pipelines at 0.16 MM$/ (″ × miles). Let's assume that the Export Pipeline length = 15 miles

Templates and manifolds are 34 × 4 wells equaling 136 MM$, and production risers are 1 × 1 × water depth (i.e. 1.5 × 5500 ft) × 10″ × 0.35/1000 equaling 28.8 MM$. Export pipelines are 0.16 × 15 miles × 8″ equaling 19.2 MM$.

12.3.7 Operating Costs (Opex)

We will use the value for tieback provided by Table 12.2 with tariff 8–13 $/boe.

So for this project (subsea tieback with tariff), we will estimate our total Opex for the field life as a minimum of 90 MMbbl × 8 and maximum of 90 MMbbl × 13 equaling 720 MM$ and 1170 MM$ for the 3 years of production on plateau.

This total cost should be prorated proportionally to production for the remaining years. For the peak rates on plateau our yearly Opex would be 75% × 720/3 equaling 180 or 75% × 1170/3 equaling 292.5 and prorated for the remaining years.

Again, a minimum of 25% Opex is recommended for the years where the production is lower than 25% of peak rate.

In the two previous case studies (cases 1 and 2 at the beginning of this chapter), we have concentrated the Opex on the plateau years. This is not totally correct. Although most of the Opex will be spent in the plateau years some will be in the remaining years.

12.3.8 Project Timing and Input to Economic Analysis

A subsea tieback will require 3 years from exploration to production on average. We can drill all four wells during this time (2 wells per year). Our project may start up production in 2014 assuming we drill our exploratory well in 2011 (Table 12.3).

Date	Oil Production rate	Gas Production rate	Exploration Drilling	Appraisal Drilling	Development Drilling	Capex	Capex	Capex	Abandonment	Opex	Comments
	stb/day	MMscf/day	MM$	MMS	MM$	MM$	MM$	MM$	MM$	MM$	
			min-max	min-max	min-max	Facility	Subsea	pipeline			
2010			55-88								1 Exploratory
2011											No appraisals
2012					(99-158) × 2		136	19.2			2 wells
2013					(99-158) × 2			28.8			2 wells
2014	28,000.00									90-146	Opex max with Tariff
2015	28,000.00									90-146	
2016	28,600.00									90-146	
2017	28,600.00									90-146	
2018	28,600.00									90-146	
2019	28,600.00									90-146	
2020	28,600.00									90-146	
2021	28,600.00									45-73	
2022	13,500.00									45-73	
2023	8,000.00									45-73	
2024	5,000.00									22-5-36.6	
2025	3,000.00									22-5-36.6	
2026	1,500.00									22-5-36.6	
2027	800.00									22-5-36.6	
2028											
2029											
2030											
2031											
2032											
Total			55-88		396-632		134	48			

Figure 12.21 Economic analysis inputs.

12.4 CASE 4—BIG ACCUMULATION >500 MMbbl IN PLACE WITH LIGHT OIL

Make an economical analysis for an oil prospect, with a volume in place of 1000 MMbbl of oil, low dissolved gas, situated in deepwater Gulf of Mexico with a water depth of 5500 ft. It is Lower Miocene age and total depth of 15,000 ft with an area of 1200 acres. Net pay will equal 200 ft. No exploratory wells, and all data is provided from seismic.

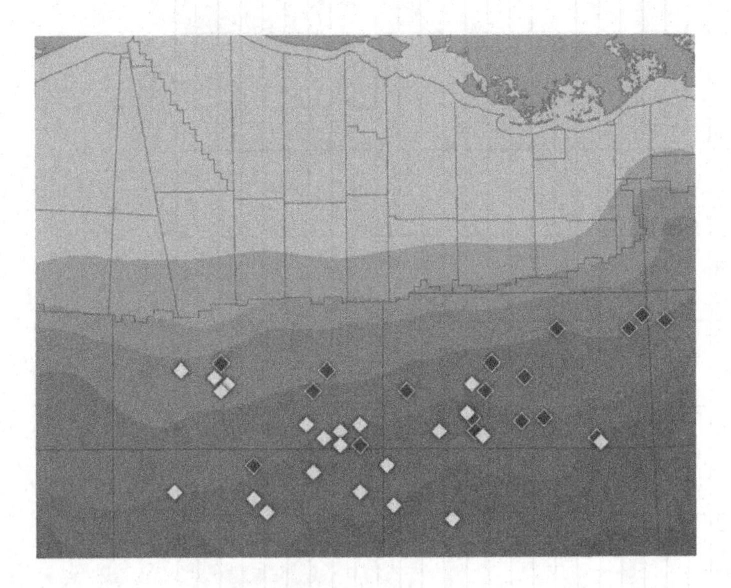

Figure 12.22 Location of a light oil prospect in deepwater GOM.

The diamonds represent potential prospects in Miocene (in gray color) and Lower Tertiary (in white).

12.4.1 Recovery Factor Estimation

Again, the first item will be to estimate the recover factor for this light oil accumulation. We can see we should expect recoveries of 19% for a lower Miocene play with reservoirs at depth of 15,000 ft. Therefore, our recoverable volume will be 19% × 1000 equaling 190 MMbbl.

12.4.2 Rock/Fluid Properties and Well Count Estimation

Next step is to derive the number of wells and their initial rate. To do this, we need first to estimate some rock and fluid properties.

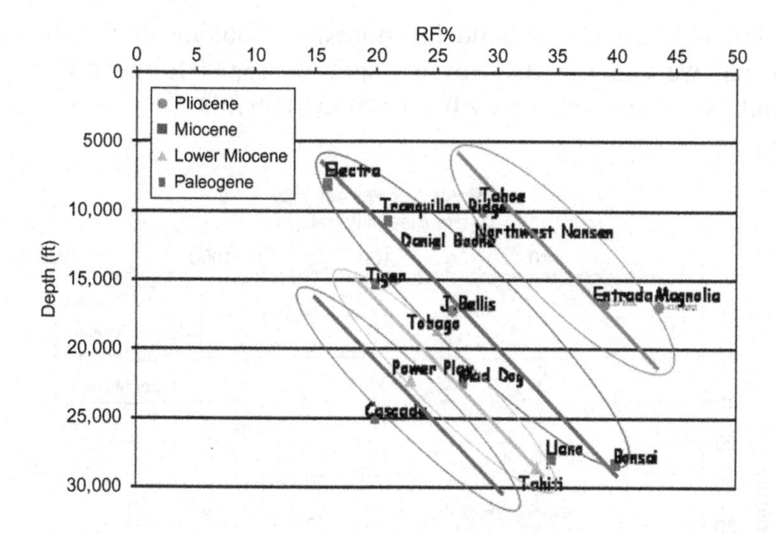

Figure 12.23 Example of a trend deriving using analogs and grouping parameters.

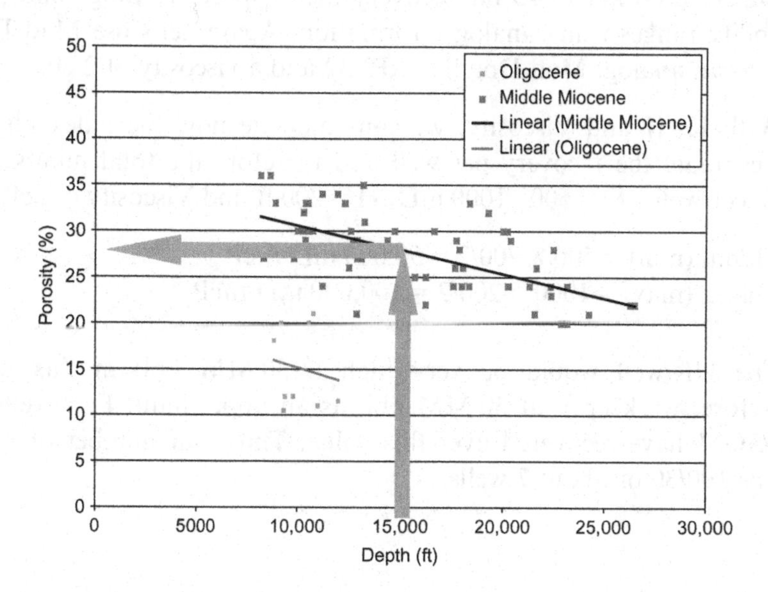

Figure 12.24 Trend for porosity vs. depth.

First, let's get an estimate for porosity. Looking at the dots for Miocene, we can see that for a depth of 15,000 ft the porosity is around 28% and permeability from 500 to 1000 mD.

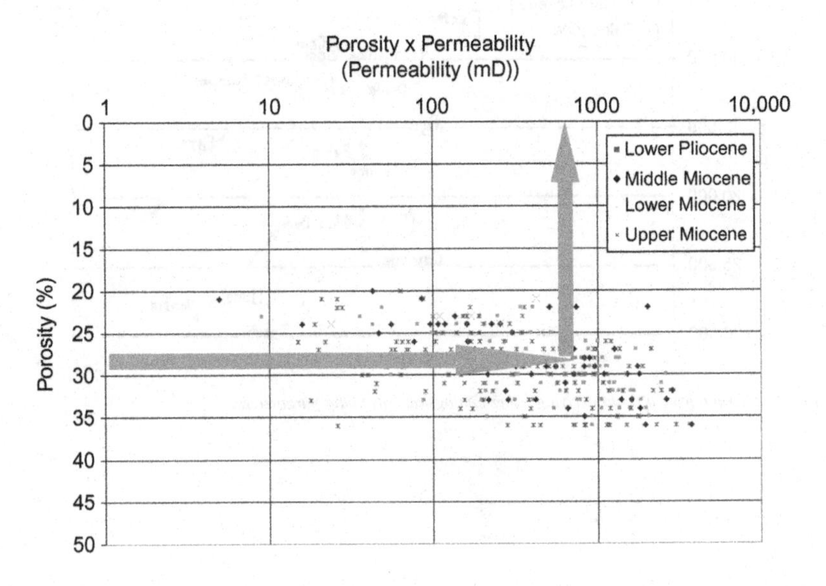

Figure 12.25 Porosity vs permeability trend.

We have estimated so far recovery factor, porosity range and permeability ranges using analog information. Again, let's use Mad Dog fluid as an analog. Mad Dog has API 32 and a viscosity of 2 cP.

With K, H and Viscosity, we can calculate now the index kh/mu and estimate the recovery per well and therefore the total number of producer wells. $K = 500–1000$ mD, $H = 200$ ft and Viscosity = 2 cP.

- Kh/mu (min) $= 500 \times 200/2 = 50{,}000$ mD ft/cP
- Kh/mu (max) $= 1000 \times 200/2 = 100{,}000$ mD ft/cP

The UR/well would be very high (>30 MM bbl) in this case. Therefore, we keep it at 30 MM bbl, as an upper limit. Few wells in the GOM have recovered over this value. Thus, our number of wells will be 190/30 or about 7 wells.

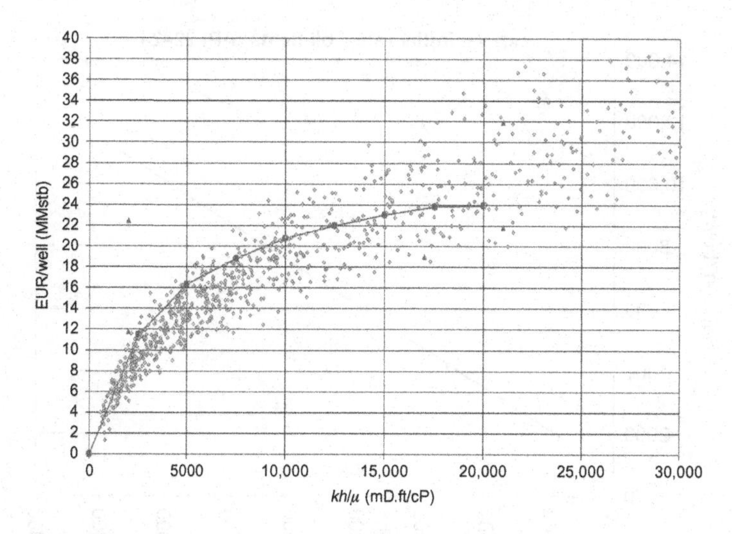

Figure 12.26 Estimation of number of wells based on rock and fluid properties.

12.4.3 Well Initial Rates and Notional Production Profile

The best way to estimate well initials is using the properties we already calculated: K (permeability), H (thickness) and Fluid Viscosity. Figure 12.27 is derived from analogs (oil fields in the GOM with API averaging around 24 API). As our fluid, we will have a higher API (32) and lower viscosity to be conservative.

Entering KH equaling 100,000, Qi will be about 7500 bbl/d. For KH equaling a 200,000 rate, Qi could be up to 12,000 bbl/d or higher. Let's use the average 10,000 bbl/d. For a fluid of 32 API these rates could be 20,000 bbl/d.

So 7 wells at 20,000 bbl/d would produce 140,000 bbl/d.

We now need to create a production curve that produces a plateau of 140,000 bbl/d or 60,000 bbl/d and total reserves of 190 MMbbl. Figure 12.28 would fit as our production scenario for the 140,000 bbl/d with 7 wells.

12.4.4 Costs: Drillex, Capex and Opex

Now it is necessary to estimate the cost to drill the 7 wells and evacuate the production as well as the operating costs.

12.4.5 Drilling Costs

Use Table 12.1 to estimate the dry hole cost.

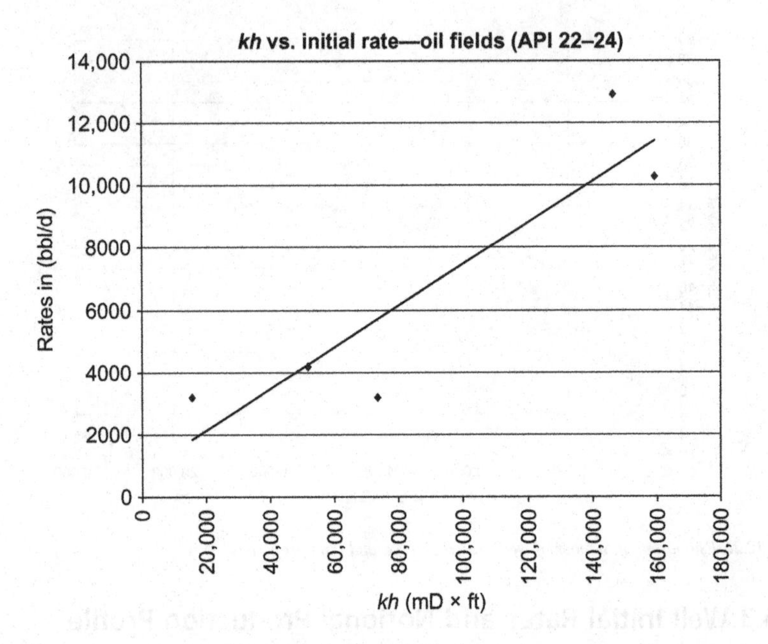

Figure 12.27 Estimation of number of wells based on rock and fluid properties.

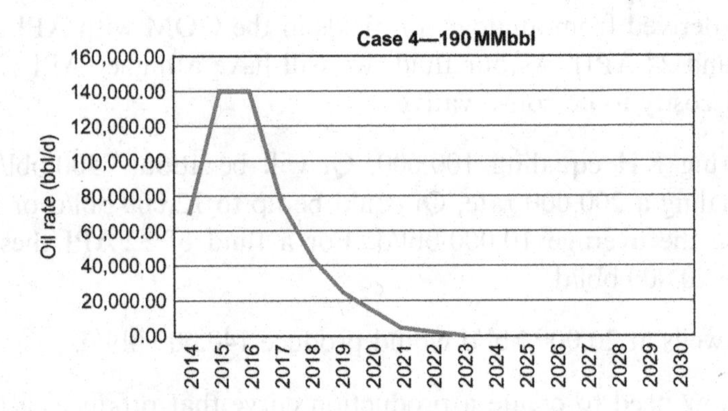

Figure 12.28 Production forecasting.

For a depth of 15,000 ft the drilling costs will be 35–64 MM$ for the exploratory well. As the volumes are between 100 MMbbl and 200 MMbbl we will need 1 appraisal well.

Therefore a development well will cost 35–64 MM$ plus completion costs (+80%) equaling 63–115 MM$. Appraisal well will cost 35–64 MM$ plus 15 days more at 500–800 MM$/day drilling rate equaling 42.5–76 MM$.

12.4.6 Facilities and Subsea Costs (Capex)

In this example, we will need a hub, but let's explore the possibility for a subsea tieback to the neighboring area where there are already 5 prospects being developed. In this case, we will only need subsea and pipeline costs. For 7 wells, we will need only one manifold plus export pipeline to the platform and the riser. Costs should look like:

- Well templates and manifolds = 34 MM$/well
- SCR Production Riser = at 0.35 MM$/(″ × 1000 ft)
- Export Pipelines at 0.16 MM$/ (″ × miles). Let's assume that the export pipeline length equals 5 miles.

Templates and manifolds are 34 × 7 wells equaling 238 MM$, and production risers are 1 × 1 × water depth (i.e. 1.5 × 5500 ft) × 10″ × 0.35/1000 equaling 28.9 MM$. Export pipelines are 0.16 × 5 miles × 8″ equaling 6.40 MM$. In this case, we have 140,000 bbl/d so we may need a bigger diameter. Therefore, our riser may go up to 5 MM$.

12.4.7 Operating Costs (Opex)

We will use the value for tieback provided by Table 12.2 with tariff 8–13 $/boe.

So for this project (subsea tieback with tariff) we will estimate our total Opex for the field life as a minimum of 190 MMbbl × 8 and maximum of 190 MMbbl × 13 equaling 1520 MM$ and 2470 MM$ for the 3 years of production on plateau.

This total cost should be prorated proportionally to production for the remaining years. For the peak rates on plateau our yearly Opex would be 50% × 1520/3 equaling 253 MM$ or 50% × 2470/3 equaling 411.66 MM$ and prorated for the remaining years.

Again, a minimum of 25% Opex is recommended for the years where the production is lower than 25% of peak rate.

12.4.8 Project Timing and Input to Economic Analysis

A subsea tieback will require 3 years from exploration to production on average. We can drill all 7 wells during this time (2 wells per year). Our project may start up production in 2015 assuming we drill our exploratory well in 2011 and the appraisal one year later in 2012.

Date	Oil Production rate	Gas Production rate	Exploration Drilling	Appraisal Drilling	Development Drilling	Capex	Capex	Capex	Abandonment	Opex	Comments
	stb/day	MMscf/day	MM$	MMS	MM$	MM$	MM$	MM$	MM$	MM$	
			min-max	min-max	min-max	Facility	Subsea	pipeline			
2010											
2011			35-64								1 Exploratory
2012				42.5-76	(63-115) × 1						1 Appraisal + 1 well
2013					(63-115) × 2		119	28.9			2 wells
2014					(63-115) × 2		119	6.4			2 wells
2015	70,000.00				(63-115) × 2					126-205	2 wells
2016	140,000.00									253-411	
2017	140,000.00									253-411	
2018	77,000.00									126-205	
2019	45,000.00									63-102	
2020	25,000.00									63-102	
2021	15,000.00									63-102	
2022	5,000.00									63-102	
2023	2,500.00									63-102	
2024	1,000.00									63-102	
2025											
2026											
2027											
2028											
2029											
2030											
2031											
2032											
Total			35-64	42.5-76	441-805		238	35.3			

Figure 12.29 Economic analysis inputs.

12.5 CASE 5—SMALL ACCUMULATION <500 bcf IN PLACE WITH DRY GAS

Make an economical analysis for a gas prospect, and with a volume in place of 500 bcf (billions of cubic feet) of dry gas. No dissolved oil, and it is situated in the deepwater Gulf of Mexico at a water depth of 6900 ft. It is Pliocene age and with a total depth of 17,000 ft with an area of 2000 acres. Net pay equals 200 ft with one exploratory well.

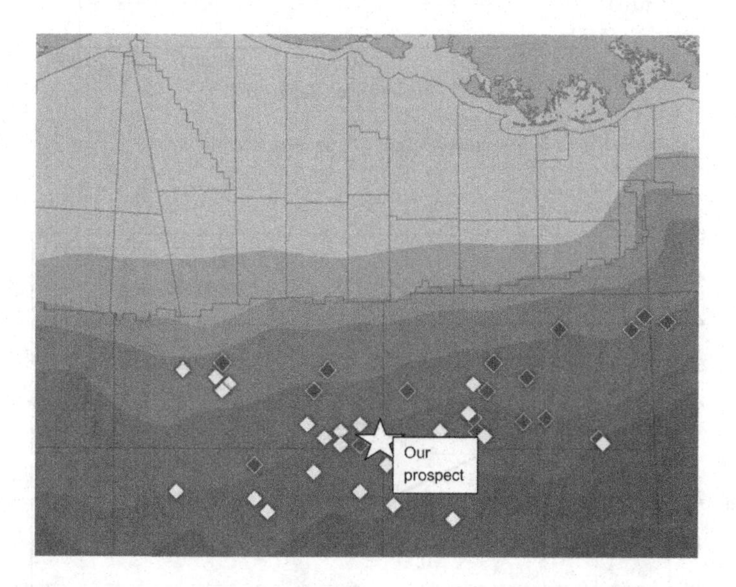

Figure 12.30 Location of a dry gas prospect, in deepwater GOM.

The diamonds represent potential prospects in Miocene (in gray color) and Lower Tertiary (in white).

12.5.1 Recovery Factor Estimation

The first item will be to estimate the recover factor but now for a gas accumulation. Figure 12.31 shows some RF analogs for gas fields (all Pliocene) in the GOM. We should expect recoveries of 60% for depth of 17,000 ft. Therefore our recoverable volume will be 60% × 500 equaling 300 bcf.

12.5.2 Rock/Fluid Properties and Well Count Estimation

Next step is to derive the number of wells and their initial rate. To do this, we need to first estimate some rock and fluid properties.

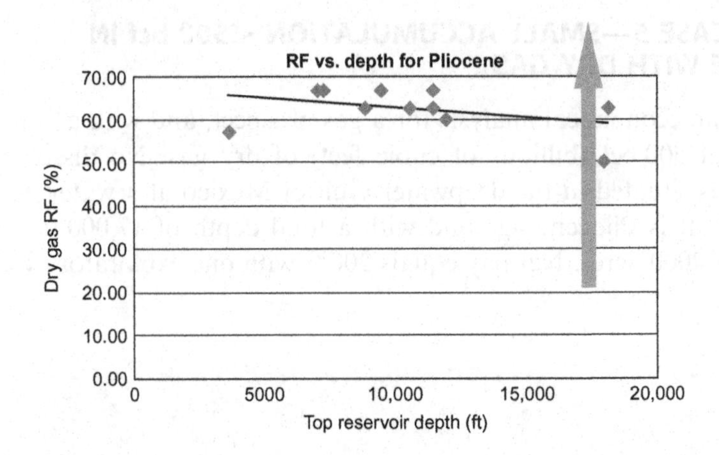

Figure 12.31 Example of a trend deriving using analogs, RF versus top reservoir depth.

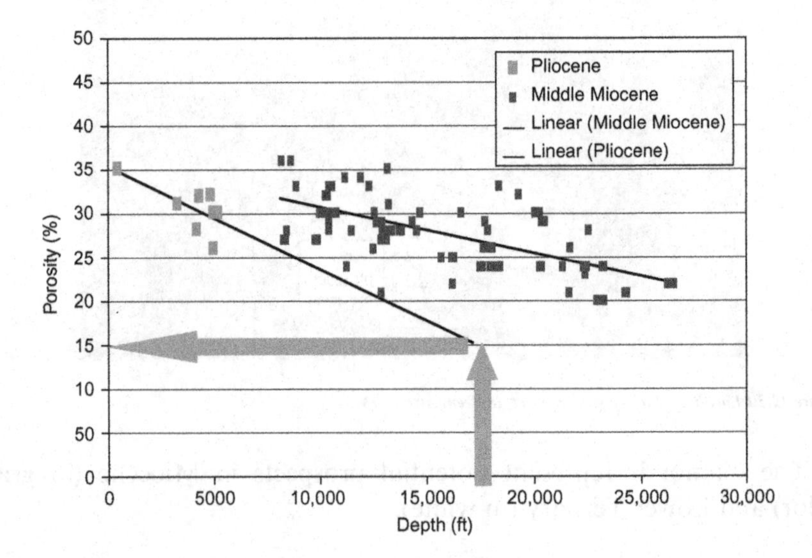

Figure 12.32 Trend for porosity vs. depth.

First, let's get an estimate for porosity. Looking at the dots for Miocene, we can see that for a depth of 17,000 ft porosity is around 15%. Our porosity × permeability trends shows porosities starting at 20% which gives a permeability range of <100 mD.

We have estimated so far recovery factor, porosity range and permeability ranges using analog information. With K, H and assuming that the gas has a very low viscosity (<0.1 cP), we can estimate the

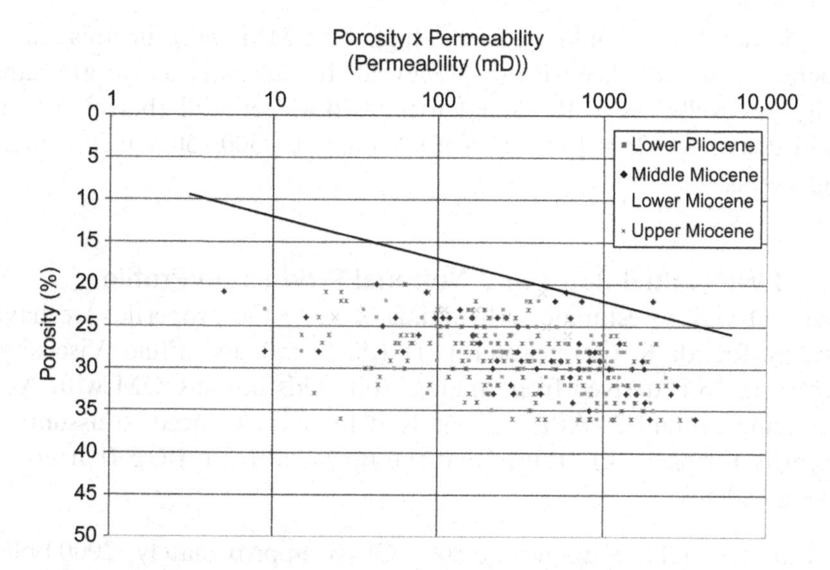

Figure 12.33 Porosity vs permeability trend.

recovery per well and therefore the total number of producer wells. K = 10–100 mD, H = 200 ft and Viscosity = 0.1 cP.

- Kh/mu (min) = 10 × 200/.1 = 20 000 mD ft/cP
- Kh/mu (max) = 100 × 200/.1 = 200 000 mD ft/cP

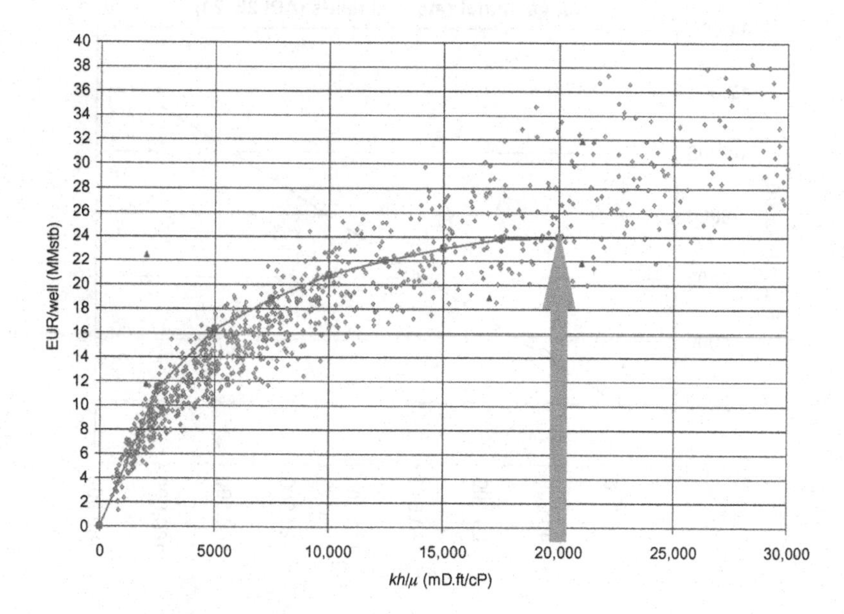

Figure 12.34 Estimation of number of wells based on rock and fluid properties.

The UR/well would be very high (>5 MM bbl) in this case. Therefore, we will keep it at 25 bbl/well. In this case, as we are handling gas wells, we will instead use 25 BOE per well (barrels of oil equivalent = 6 × 25 = 150 bcf/ well) leading to 300/150 which equals 2 gas wells.

12.5.3 Well Initial Rates and Notional Production Profile

The best way to estimate well initials is using the properties we have already found: K (permeability), H (thickness) and Fluid Viscosity. Figure 12.35 is derived from analogs (oil fields in the GOM with API averaging around 24 API). To apply it for gas, we need to assume a viscosity for gas = 0.1 cP and that the rates will be in BOE (barrels of oil equivalent).

Entering KH equaling 20,000, Qi is approximately 2000 bbl/d equaling 12 MMscf/d. KH equaling a 200,000 rate, Qi could be up to 12,000 bbl/d equaling 72 MMscf/d. Let's use the average 10,000 bbl/d equaling 60 MMscf/d.

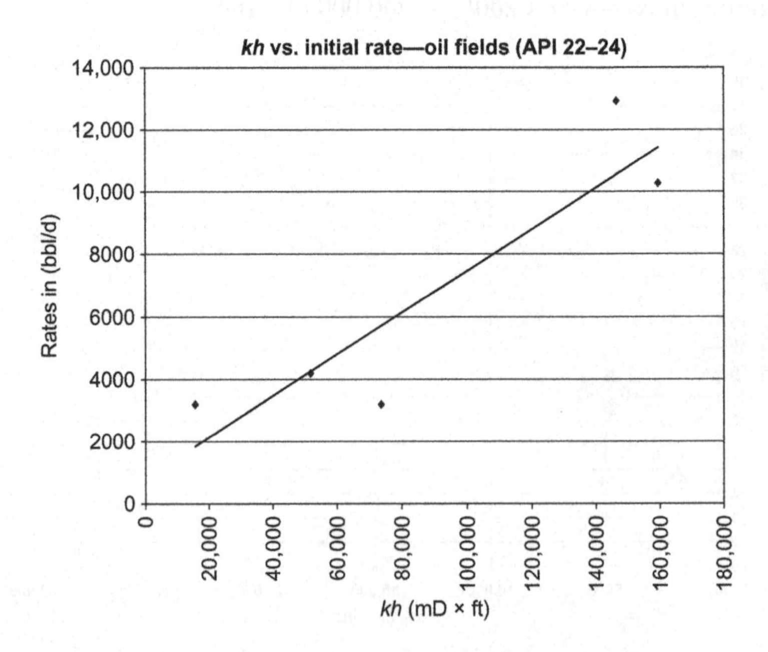

Figure 12.35 Estimation of well rates based on rock and fluid properties.

So with 2 wells at 60 MMscf/d, we would produce a plateau of 120 MMscf/d. Our production curve will look like the one shown in Figure 12.36 for 120 MMscf/d with 2 wells.

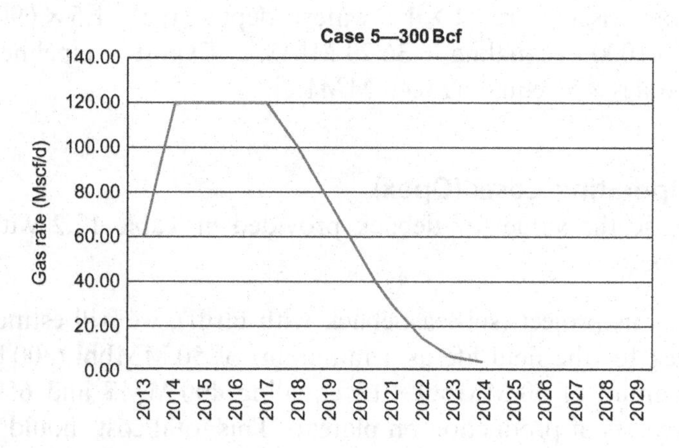

Figure 12.36 Production forecasting.

12.5.4 Costs: Drillex, Capex and Opex

Now it is necessary to estimate the cost to drill the 2 wells and evacuate the production as well as the operating costs.

12.5.5 Drilling Costs

Use Table 12.1 to estimate the dry hole cost.

For a depth of 17,000 ft, the drilling costs will be 35–64 MM$ for the exploratory well. As the volumes are small—less than 100 MMbbl (300 Bcf)—we will not need any appraisal well. Therefore, a development well will cost 35–64 MM$ plus completion costs (+80%) equaling 63–115 MM$.

12.5.6 Facilities and Subsea Costs (Capex)

In this study, we may need a hub, but let's explore the possibility for a subsea tieback to the neighboring area where there are already several prospects being developed. In this case, we will only need subsea and pipeline costs. For 2 wells, we will need only one manifold plus export pipeline to the platform and the riser. The costs should look like:

- Well templates and manifolds = 34 MM$/well
- SCR Production Riser = at 0.35 MM$/(″ × 1000 ft)

- Export Pipelines at 0.16 MM$/ (″ × miles). Let's assume that the export pipeline length equals 5 miles

Templates and manifolds are 34 × 2 wells equaling 68 MM$, and production risers are 1 × 1 × water depth (i.e. 1.5 × 6900 ft) × 10″ × 0.35/1000 equaling 36.23 MM$. Export pipelines are 0.16 × 5 miles × 8″ equaling 6.40 MM$.

12.5.7 Operating Costs (Opex)

We will use the value for tieback provided in Table 12.2 with tariff 8–13 $/boe.

So for this project (subsea tieback with tariff), we will estimate our total Opex for the field life as a minimum of 50 MMbbl (300 bcf) × 8 and maximum of 50 MMbbl × 13 equaling 400 MM$ and 650 MM$ for the 4 years of production on plateau. This total cost should be prorated proportionally to production for the remaining years.

For the peak rates on plateau, our yearly Opex would be 50% × 400/4 equaling 50 MM$ or 50% × 640/4 equaling 80 MM$ and prorated for the remaining years. Again, the Opex distribution during the years is merely indicative and should be based on real operating costs for current and previous projects.

12.5.8 Project Timing and Input to Economic Analysis

A subsea tieback will require 3 years from exploration to production on average. We can drill the 2 wells in 2011. Our project may start up production in 2013 assuming we have already drilled our exploratory well in 2010.

Date	Oil Production rate	Gas Production rate	Exploration Drilling	Appraisal Drilling	Development Drilling	Capex	Capex	Capex	Abandonment	Opex	Comments
	stb/day	MMscf/day	MM$	MMS	MM$	MM$	MM$	MM$	MM$	MM$	
			min-max	min-max	min-max	Facility	Subsea	pipeline			
2010			35-64								1 Exploratory
2011					(63-115) × 2		34	36.23			2 wells
2012							34	6.4			
2013		60.00								25-40	
2014		120.00								50-80	
2015		120.00								50-80	
2016		120.00								50-80	
2017		120.00								50-80	
2018		100.00								25-40	
2019		75.00								25-40	
2020		51.00								25-40	
2021		30.00								12.5-20	
2022		15.00								12.5-20	
2023		6.00								12.5-20	
2024		3.00								12.5-20	
2025											
2026											
2027											
2028											
2029											
2030											
2031											
2032											
Total			35-64		126-230		68	42.63			

Figure 12.37 Economic analysis inputs

12.6 CASE 6—BIG ACCUMULATION >500 bcf IN PLACE WITH DRY GAS

This example will be similar to case 5, but the only difference it is now for a bigger volume (1 Tcf equals 1000 Bcf). All other parameters will be the same, except the area is now 4000 acres. Net pay will equal 200 ft.

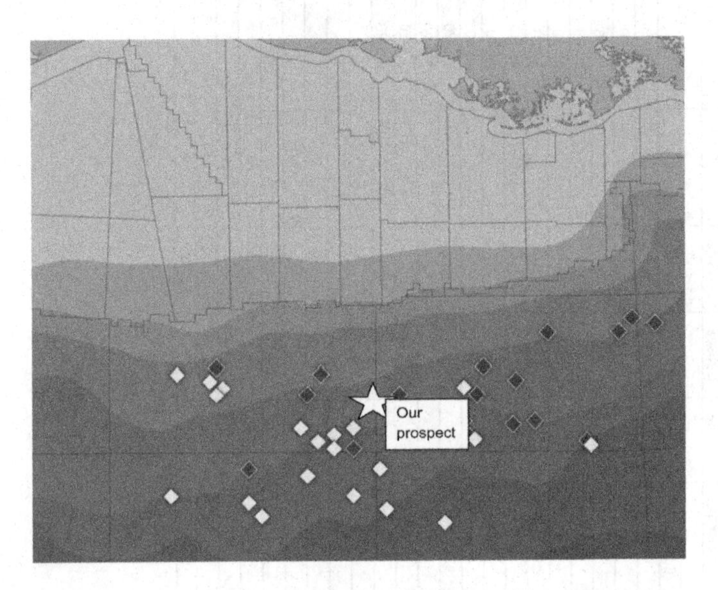

Figure 12.38 Location of a dry gas prospect, in deepwater GOM.

The diamonds represent potential prospects in Miocene (in gray color) and Lower Tertiary (in white).

12.6.1 Recovery Factor Estimation

The first thing will be to estimate the recover factor. Figure 12.39 shows some RF analogs for gas fields (all Pliocene) in the GOM. We should expect recoveries of 60% for depth of 17,000 ft. Therefore, our recoverable volume will be 60% × 1000 equaling 600 bcf.

12.6.2 Rock/Fluid Properties and Well Count Estimation

Next step is to derive the number of wells and their initial rate. To do this, we need first to estimate some rock and fluid properties.

First, let's get an estimate for porosity. Looking at the dots for Miocene, we can see that for a depth of 17,000 ft porosity is around

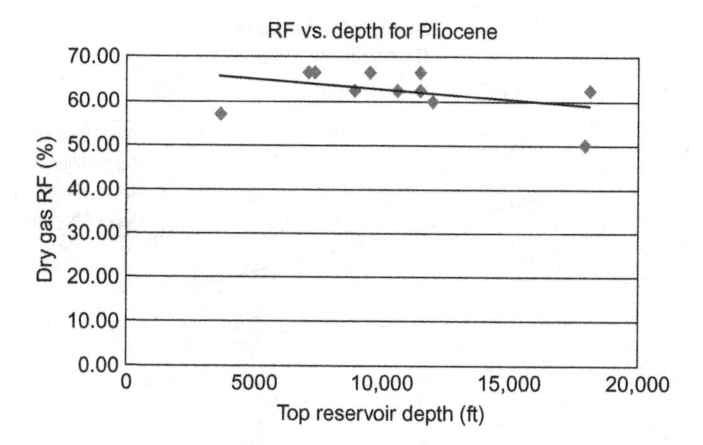

Figure 12.39 Example of a trend deriving using analogs, RF versus top reservoir depth.

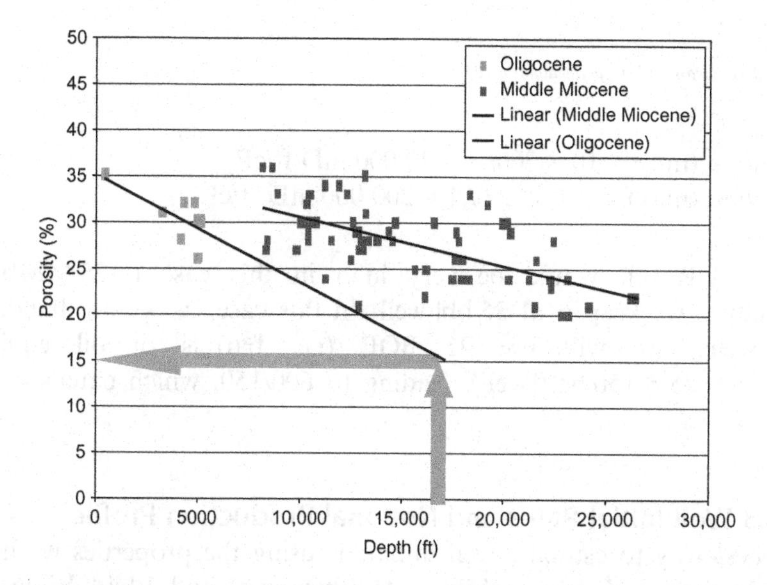

Figure 12.40 Trend for porosity vs. depth.

15%. Our porosity × permeability trends shows porosities starting at 20% which gives a permeability range <100 mD.

We have estimated so far recovery factor, porosity range and permeability ranges using analog information. With K, H, and assuming that the gas has a very low viscosity (i.e. <0.1 cP), we can estimate the recovery per well and therefore the total number of producer wells. K = 10–100 mD, H = 200 ft and Viscosity = 0.1 cP.

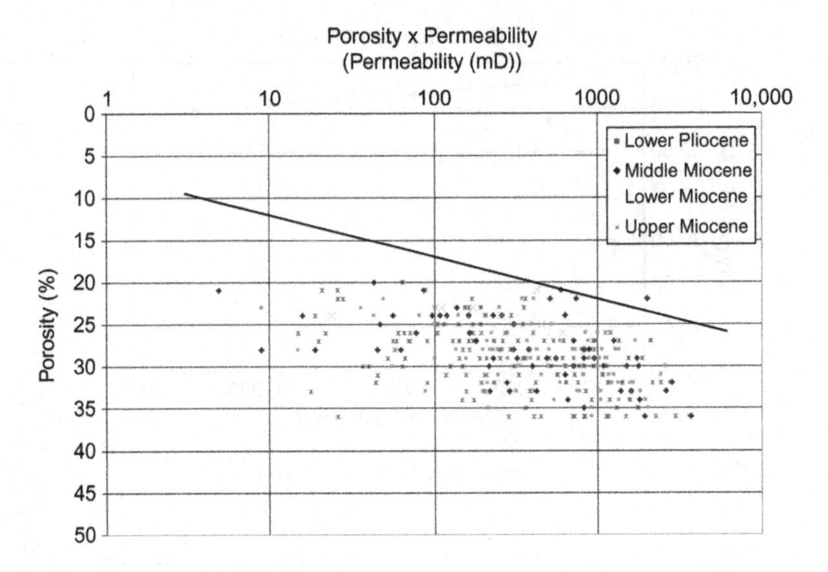

Figure 12.41 Porosity versus permeability trend.

- Kh/mu (min) = 10 × 200/.1 = 20 000 mD ft/cP
- Kh/mu (max) = 100 × 200/.1 = 200 000 mD ft/cP

The UR/well would be very high in this case (>25 MMbbl). Therefore, we keep it at 25 bbl/well. In this case, as we are handling gas wells, we will use 25 BOE (i.e. barrels of oil equivalent = 6 × 25 = 150 bcf/ well), leading to 600/150, which equals 4 gas wells.

12.6.3 Well Initial Rates and Notional Production Profile

The best way to estimate well initials is using the properties we have already found: K (permeability), H (thickness) and Fluid Viscosity. Figure 12.42 is derived from analogs (oil fields in the GOM with API averaging around 24 API). To apply it for gas, we need to assume a viscosity for gas = 0.1 cP and that the rates will be in BOE (barrels of oil equivalent).

KH equals 20,000, making Qi approximately 2000 bbl/d equaling 12 MMscf/d. KH equals a 200,000 rate, meaning Qi could go up to 12,000 bbl/d equaling 72 MMscf/d. Let's use the average 10,000 bbl/d which equals 60 MMscf/d.

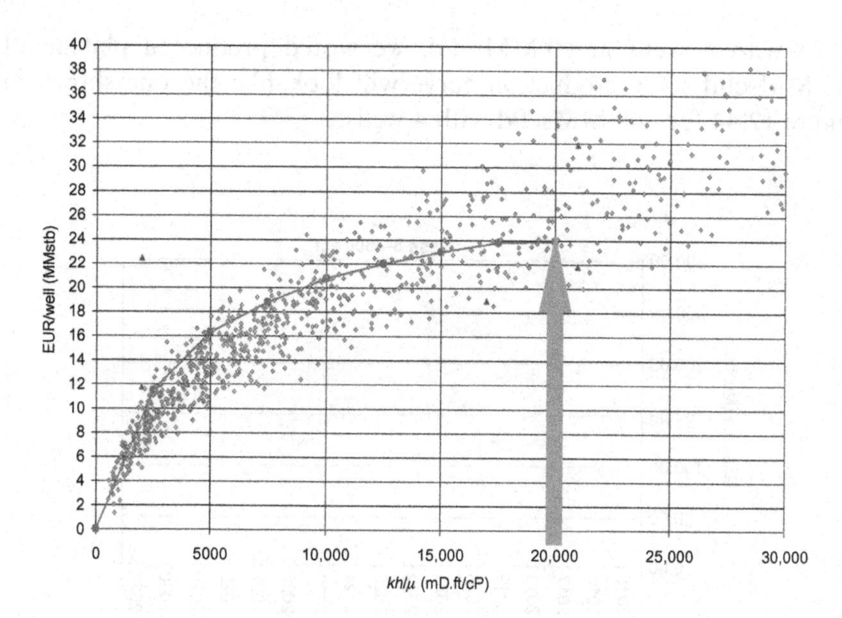

Figure 12.42 Estimation of number of wells based on rock and fluid properties.

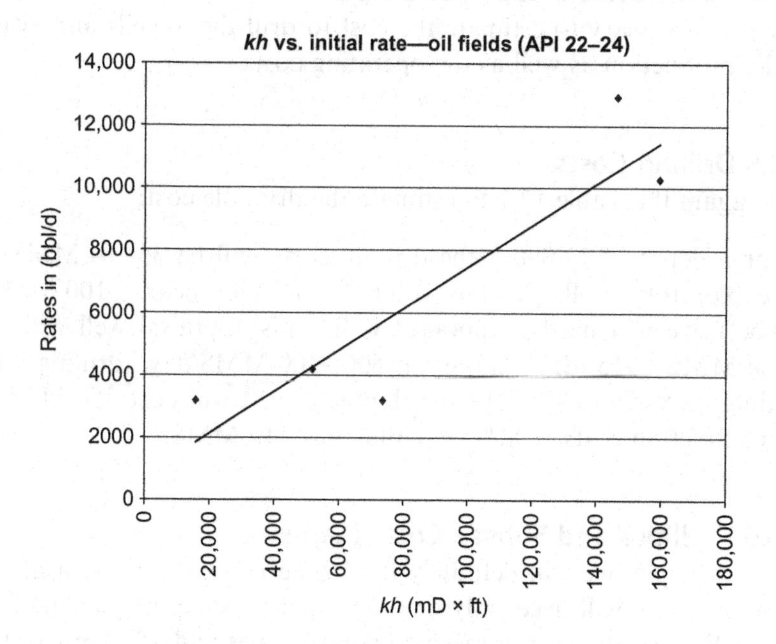

Figure 12.43 Estimation of well rates based on rock and fluid properties.

So with 4 wells at 60 MMscf/d, we would produce a plateau of 240 MMscf/d. Our production curve will look like the one shown in Figure 12.44 for 240 MMscf/d with 4 wells.

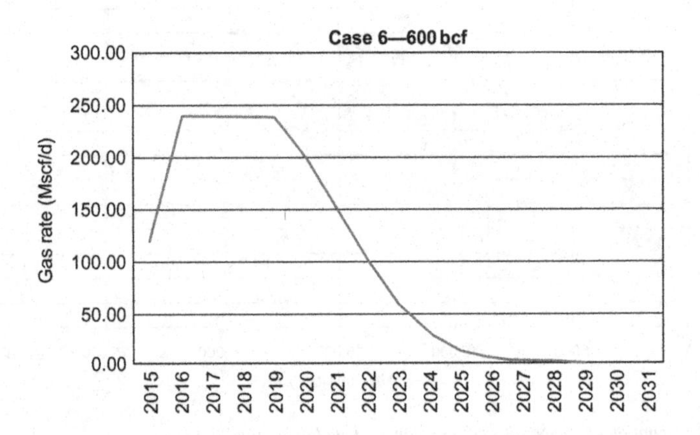

Figure 12.44 Production forecasting.

12.6.4 Costs: Drillex, Capex and Opex

Now it is necessary to estimate the cost to drill the 4 wells and evacuate the production as well as the operating costs.

12.6.5 Drilling Costs

Using again the Table 12.1 to estimate the dry hole cost:

For a depth of 17,000 ft the drilling costs will be 35–64 MM$ for the exploratory well. As the volumes are big, nearly 100 MMbbl (600 bcf), we will need 1 appraisal well. This appraisal well will cost 35–64 MM$ + (15 days more at 500–800 MM$/day, drilling rate) equaling 42.5–76 MM$. The development well will cost 35–64 MM$ plus completion costs (+ 80%) equaling 63–115 MM$.

12.6.6 Facilities and Subsea Costs (Capex)

In this example, we will definitely need a hub so we will use a SPAR. For 4 wells, we will need only one manifold plus export pipeline from the SPAR to an existing infrastructure (100 miles and 25″ diameter).

Figure 12.45 will be used again to provide an estimation of Capex for a SPAR of capacity 250 MMscf/d (about 50 kbbl/d) equaling $ MM 400 for Hull.

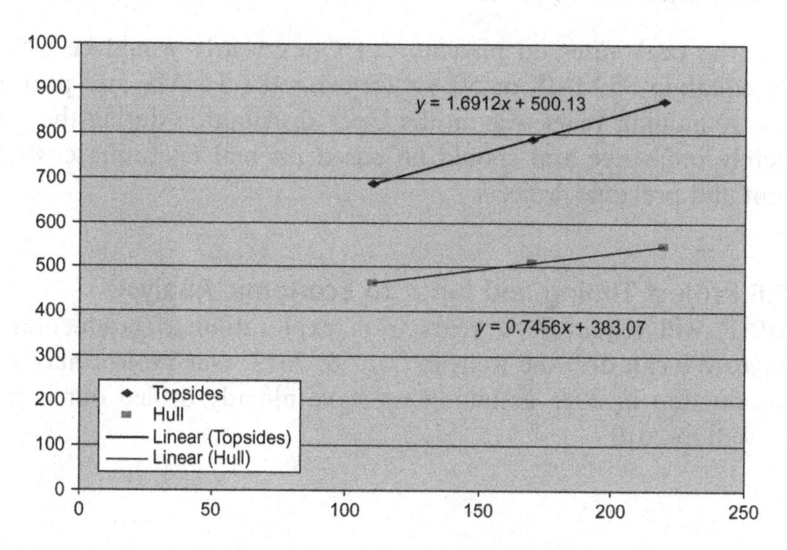

Figure 12.45 Hull and topsides Capex estimation MM$ versus FPU capacity.

And for the subsea:

- Well templates and manifolds = 34 MM$/well
- SCR Production Riser (diameter = 8″) = at 0.35 MM$/(″ × 1000 ft)
- Export Pipelines at 0.16 MM$/ (″ × miles). Let's assume that the Export Pipeline Length = 100 miles and 25″ diameter
- Intra-field flow lines at 0.22 MM$/(″ × miles) at 8 inches and 5 miles

Templates and manifolds are 34 × 4 wells equaling 136 MM$, and production risers are 1 × 1 × water depth (i.e. 1.5 × 6900 ft) × 8″ × 0.35/1000 equaling 28.98 MM$. Export pipelines are 0.16 × 100 miles × 25″ equaling 400 MM$, and intra-field flow lines are 0.22 × 5 miles × 8 equaling 8.8 MM$.

12.6.7 Operating Costs (Opex)

We will use the value for Standalone FPU (SPAR) provided by Table 12.2 with no tariff 8–15 $/BOE as we will take care of building our own pipeline for the gas evacuation.

So for this project, we will estimate our total Opex for the field life as a minimum of 100 MMbbl (600 bcf) × 8 and maximum of 100 MMbbl × 15 equaling 1500 MM$. This total cost should be prorated proportionally to production for the remaining years.

For the peak rates on plateau, our Opex yearly would be 50% × 600/4 equaling 75 MM$ or 50% × 1500/4 = 187.5 MM$ and prorated for the remaining years. Again, the Opex distribution during the years is merely indicative and should be based on real operating costs for current and previous projects.

12.6.8 Project Timing and Input to Economic Analysis

A SPAR will require 4–5 years from exploration to production on average. We can drill the wells in 2012 & 2013. Our project may start up production in 2015 assuming we have already drilled our exploratory well in 2010.

Date	Oil Production rate	Gas Production rate	Exploration Drilling	Appraisal Drilling	Development Drilling	Capex	Capex	Capex	Abandonment	Opex	Comments
	stb/day	MMscf/day	MM$	MMS	MM$	MM$	MM$	MM$	MM$	MM$	
			min-max	min-max	min-max	Facility	Subsea	pipeline			
2010			35-64								1 Exploratory
2011				42.6-76			34				1 Appraisal
2012					(63-115) × 2	200	34	8.8			2 wells
2013					(63-115) × 2	200	34	28.98			2 wells
2014							34	400			
2015		120.00								37.5-93.75	
2016		240.00								75-187.5	
2017		240.00								75-187.5	
2018		240.00								75-187.5	
2019		240.00								75-187.5	
2020		200.00								75-187.5	
2021		150.00								37.5-93.75	
2022		102.00								37.5-93.75	
2023		60.00								18.25-46.875	
2024		30.00								9.12-23.4	
2025		12.00								9.12-23.4	
2026		6.00								9.12-23.4	
2027		3.00								9.12-23.4	
2028		2.00								9.12-23.4	
2029		1.00								9.12-23.4	
2030											
2031											
2032											
Total			35-64	42.6-76	252-460	400	136	437.78			

Figure 12.46 Economic analysis inputs.

12.7 CASE 7—SMALL ACCUMULATION <500 bcf, GAS AND CONDENSATE

This example will be a gas condensate prospect similar to case study 5 (500 bcf). All other parameters will be the same, except that instead of dry gas now we have a gas condensate with CGR equaling 10 bbl/scf.

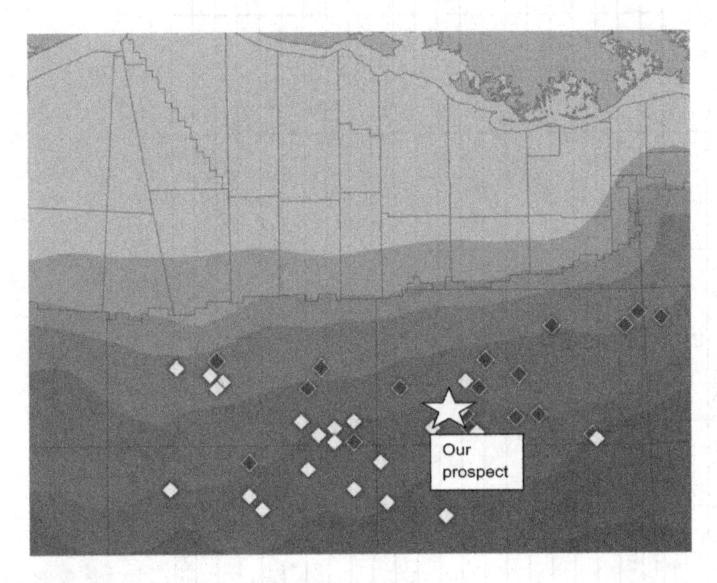

Figure 12.47 Location of a condensate gas prospect in deepwater GOM.

The diamonds represent potential prospects in Miocene (in gray color) and Lower Tertiary (in white).

12.7.1 Recovery Factor Estimation

Let's estimate the recover factor. Figure 12.48 shows some RF analogs for gas condensate fields (all ages) in the GOM as a function of CGR (condensate gas ratio). For a CGR = 10 bbl/scf, our recoverable volume will be 55% × 500 = 275 bcf. Looking at some analog fields in the Gulf of Mexico at the reservoir depth of 17,000 ft, we can see that for a depth of 17,000 ft the RF would be slightly less, nearly 52% although some analogs would show 48%, as shown in Figure 12.49. This helps us identify a trend of gas condensate fields versus depth.

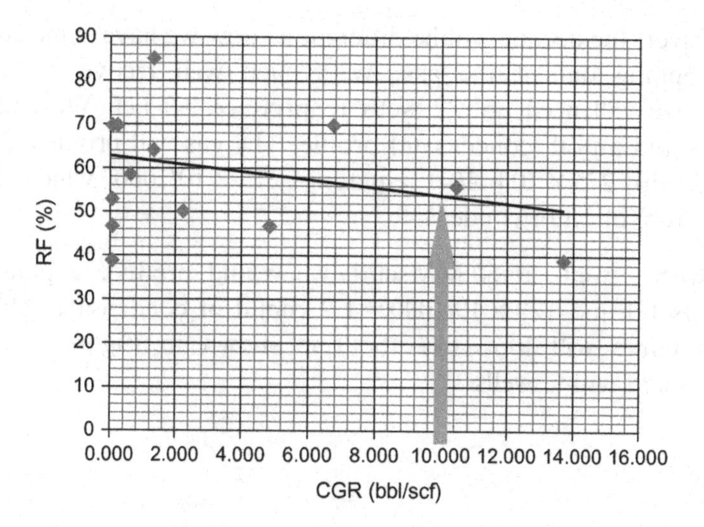

Figure 12.48 Example of a trend deriving using analogs, RF versus CGR.

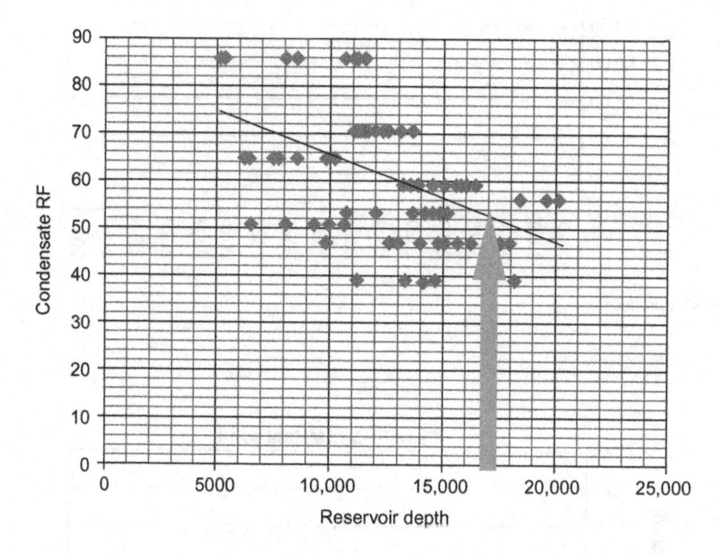

Figure 12.49 Example RF for gas condensate versus reservoir depth.

12.7.2 Rock/Fluid Properties and Well Count Estimation

This would be the same as in the previous study as we are changing the fluid here and not the rock properties. The recovery per well and initial rate would be the same as the previous case.

However, the decline will be different as now we have some condensate dropping out of the gas. With each well recovering around 150 bcf, we will need only 2 wells to produce 275 bcf. We will have now gas and an oil (condensate) profile. The gas will produce 275 bcf and the oil 275 × 10 bbl/scf equaling 275 × 10/1000 which equals 2.75 MMbbl of oil or condensate.

So with 2 wells at 60 MMscf/d, we would produce a plateau of 120 MMscf/d and 120 × 10/1000 = 1.2 kbbl/d of condensate. The production curve will look like the one shown in Figure 12.50 for 120 MMscf/d with 2 wells.

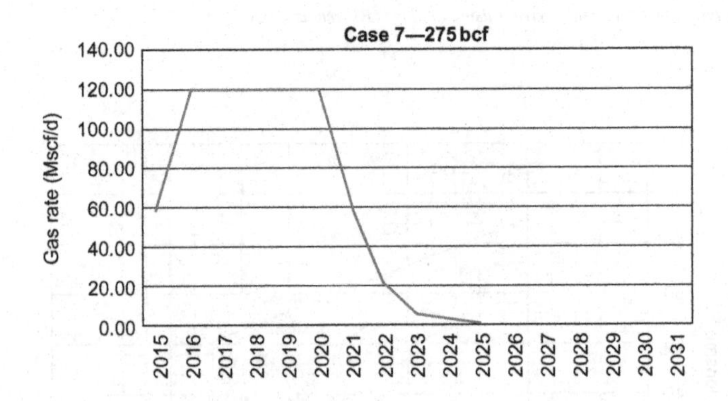

Figure 12.50 Gas production.

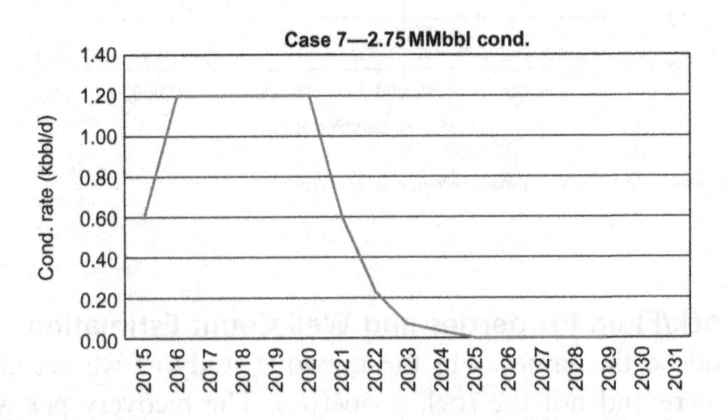

Figure 12.51 Condensate production.

12.7.3 Costs: Drillex, Capex and Opex

Now it is necessary to estimate the cost to drill the 2 wells and evacuate the production as well as the operating costs.

12.7.4 Drilling costs

Use again Table 12.1 to estimate the dry hole cost.

For a depth of 17,000 ft the drilling costs will be 35–64 MM$ for the exploratory well. As the volumes are now smaller again, there is no need for an appraisal well. The development well will cost 35–64 MM$ + completion costs (+80%) = 63–115 MM$.

12.7.5 Facilities and Subsea Costs (Capex)

In this example, we will use the same SPAR as in the previous case study, but now we will have both gas and condensate export lines (gas and condensate to be separated in the SPAR) to an existing hub. For 2 wells, we will need only one manifold plus export pipeline (gas and condensate) to an existing infrastructure. Costs should look like:

- Well templates and manifolds = 34 MM$/well
- SCR Production Riser (diameter = 8″) = at 0.35 MM$/(″ × 1000 ft)
- 2 × Export Pipelines at 0.16 MM$/ (″ × miles). Let's assume that the export pipeline length = 100 miles and 16″ diameter
- Intra-field flow lines at 0.22 MM$/(″ × miles) at 8 inches and 5 miles

Templates and manifolds are 34 × 2 wells equaling 68 MM$, and production risers are 1 × 1 × water depth (i.e. 1.5 × 6900 ft) × 8″ × 0.35/1000 equaling 28.98 MM$. Export pipelines × 2 are 0.16 × 100 miles × 16″ equaling 512 MM$, and intra-field flow lines are 0.22 × 5 miles × 8 equaling 8.8 MM$.

12.7.6 Operating Costs (Opex)

We will use the value for Standalone FPU (SPAR) provided by Table 12.2 with no tariff (i.e. 8–15 $/boe) as we will take care of building our own pipeline for the gas evacuation.

So for this project, we will estimate our total Opex for the field life as a minimum of 45.83 MMbbl (275 bcf) × 8 equaling 366.64 and maximum of 45.83 MMbbl × 15 equaling 687.45 MM$.

This total cost should be prorated proportionally to production for the remaining years.

For the peak rates on plateau, our Opex yearly would be 50% × 366.64/5 equaling 36.664 MM$ or 50% × 687.45/5 equaling 68.745 MM$ and prorated for the remaining years. Again, the Opex distribution during the years is merely indicative and should be based on real operating costs for current and previous projects.

12.7.7 Project Timing and Input to Economic Analysis

A SPAR will require 4–5 years from exploration to production on average. We can drill the wells in 2012 & 2013. Our project may start up production in 2015 assuming we have already drilled our exploratory well in 2010.

Date	Oil Production rate	Gas Production rate	Exploration Drilling	Appraisal Drilling	Development Drilling	Capex	Capex	Capex	Abandonment	Opex	Comments
	stb/day	MMscf/day	MM$	MMS	MM$	MM$	MM$	MM$	MM$	MM$	
			min-max	min-max	min-max	Facility	Subsea	pipeline			
2010			35-64								1 Exploratory
2011											No appraisals well
2012					(63-115) × 1	200	136	8.8			1 well
2013					(99-115) × 1	200		28.98			1 well
2014							68	512			
2015	1.20	120.00								18.33-34.37	
2016	2.40	240.00								36.66.68.74	
2017	2.40	240.00								36.66.68.74	
2018	2.40	240.00								36.66.68.74	
2019	2.40	240.00								36.66.68.74	
2020	2.00	200.00								36.66.68.74	
2021	1.50	150.00								18.33-34.37	
2022	1.02	102.00								18.33-34.37	
2023	0.60	60.00								18.33-34.37	
2024	0.30	30.00								9.17-17.19	
2025	0.12	12.00								9.17-17.19	
2026	0.06	6.00								9.17-17.19	
2027	0.03	3.00								9.17-17.19	
2028	0.02	2.00								9.17-17.19	
2029	0.02	1.00								9.17-17.19	
2030										9.17-17.19	
2031											
2032											
Total			35-64		126-230	400	68	549.78			

Figure 12.52 Economic analysis inputs.

12.8 CASE 8—BIG ACCUMULATION >500 bcf, GAS AND CONDENSATE

This case study will be a gas condensate prospect similar to case study 6 (1 Tcf). All other parameters will be the same, except that instead of dry gas now we have a gas condensate with CGR = 10 bbl/scf.

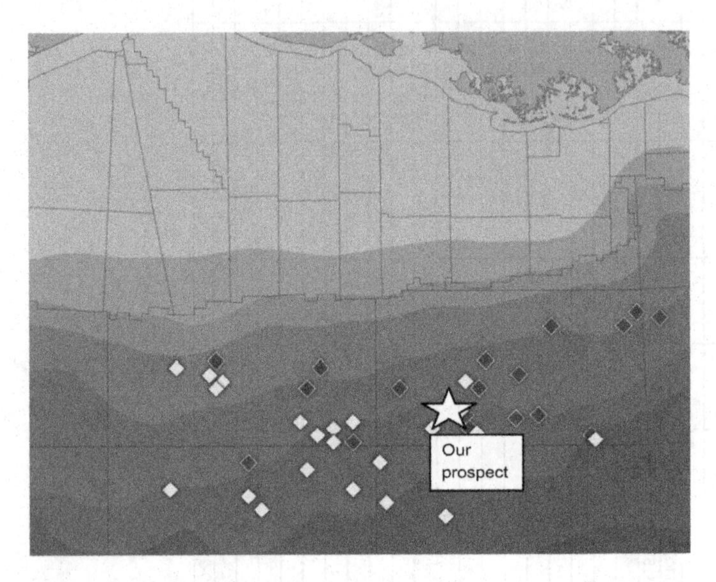

Figure 12.53 Location of a condensate gas prospect in deepwater GOM.

The diamonds represent potential prospects in Miocene (in gray color) and Lower Tertiary (in white).

12.8.1 Recovery Factor Estimation

Let's estimate the recover factor. Figure 12.54 shows some RF analogs for gas condensate fields (all ages) in the GOM as a function of CGR (condensate gas ratio). For a CGR = 10 bbl/scf, our recoverable volume will be 55% × 1000 equaling 550 bcf. Looking at more analog fields at the reservoir depth of 17,000 ft, at the trend of gas condensate fields in the Gulf of Mexico versus depth, we can see that for a depth of 17,000 ft the RF would be slightly less, nearly 52%, although some analogs would show 48%.

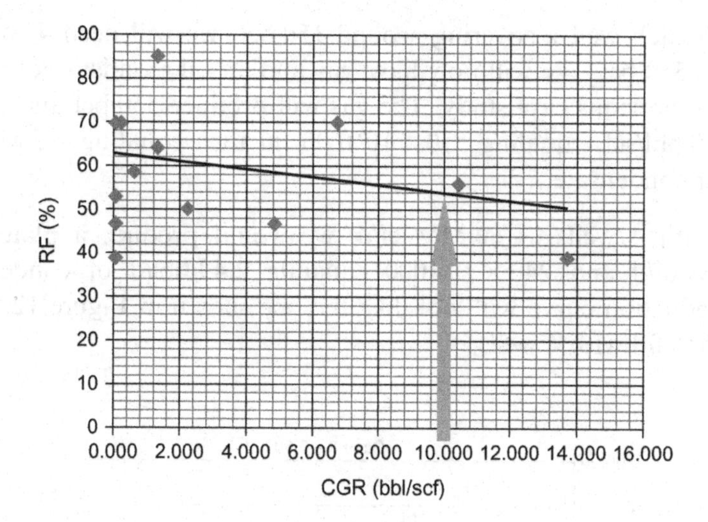

Figure 12.54 Example of a trend deriving using analogs, RF versus CGR.

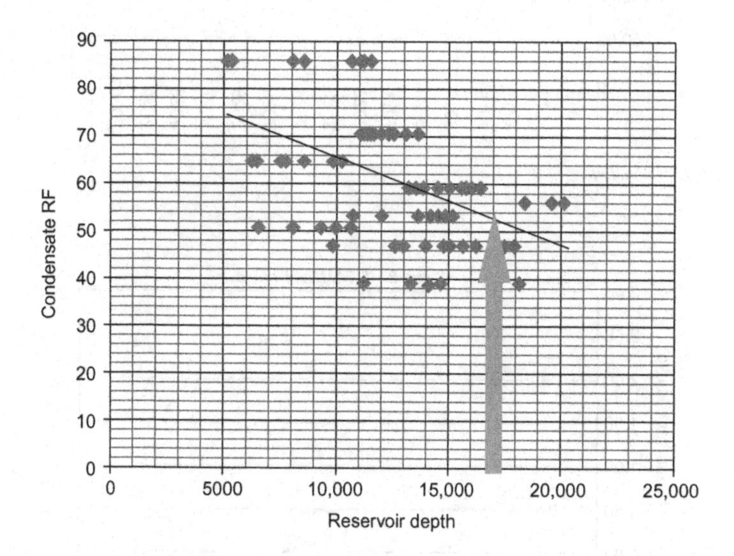

Figure 12.55 Example RF for gas condensate versus reservoir depth.

12.8.2 Rock/Fluid Properties and Well Count Estimation

This would be the same as in the previous cast study as we are changing the fluid here and not the rock properties. The recovery per well and initial rate would be the same as for the previous case.

With each well recovering around 150 bcf, we will need 4 wells to produce 550 bcf. We will now have gas and an oil (condensate) profile as in the previous case study. The gas will produce 550 bcf and the oil 550 × 10 bbl/scf, equaling 550 × 10/1000, in turn equaling 5.5 MMbbl of oil or condensate.

So with 4 wells at 60 MMscf/d, we would produce a plateau of 240 MMscf/d and 240 × 10/1000 equaling 2.4 kbbl/d of condensate. The production curve will look like the one shown in Figure 12.56 for 240 MMscf/d with 4 wells.

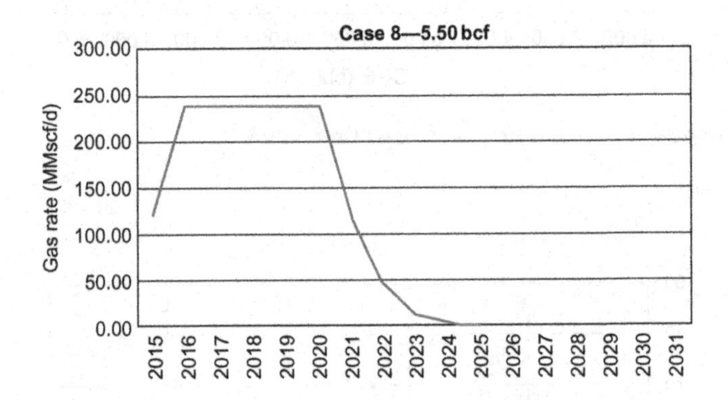

Figure 12.56 Gas production.

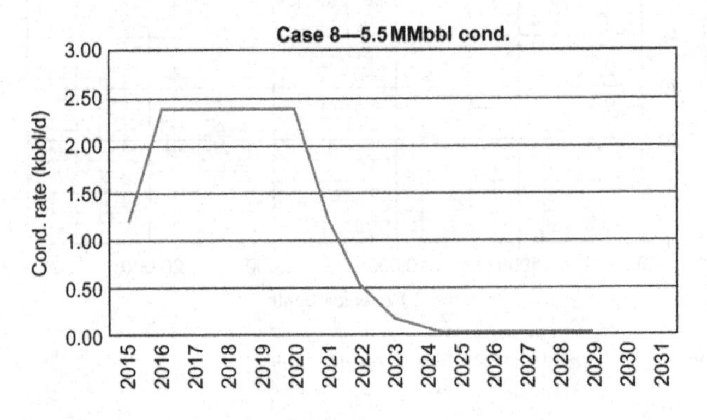

Figure 12.57 Condensate production.

12.8.3 Costs: Drillex, Capex and Opex

Now it is necessary to estimate the cost to drill the 2 wells, and evacuate the production as well as the operating costs.

12.8.4 Drilling costs

Use Table 12.1 to estimate the dry hole cost.

For a depth of 17,000 ft the drilling costs will be 35–64 MM$ for the exploratory well. As the volumes are now smaller again, there is no need for an appraisal well. The development well will cost 35–64 MM$ plus completion costs (+ 80%) equaling 63–115 MM$.

12.8.5 Facilities and Subsea Costs (Capex)

In this example, we will use the same SPAR as in the previous case study, but now we will have both gas and condensate export lines (gas and condensate to be separated in the SPAR) sent to an existing hub. For 2 wells, we will need only one manifold plus export pipeline (gas and condensate) to an existing infrastructure. Costs should look like:

- Well templates and manifolds = 34 MM$/well
- SCR Production Riser (diameter = 8″) = at 0.35 MM$/(″ × 1000 ft)
- 2 × Export Pipelines at 0.16 MM$/ (″ × miles). Let's assume that the export pipeline length = 100 miles and 16″ diameter
- Intra-field flow lines at 0.22 MM$/(″ × miles) at 8 inches and 5 miles

Templates and manifolds are 34 × 2 wells equaling 68 MM$, and production risers are 1 × 1 × water depth (i.e. 1.5 × 6900 ft) × 8″ × 0.35/1000 equaling 28.98 MM$. Export pipelines × 2 are 0.16 × 100 miles × 16″ equaling 512 MM$, and intra-field flow lines are 0.22 × 5 miles × 8 equaling 8.8 MM$.

12.8.6 Operating Costs (Opex)

We will use the value for Standalone FPU (SPAR) provided by Table 12.2 with no tariff (i.e. 8–15 $/boe) as we will take care of building our own pipeline for the gas evacuation.

So for this project we will estimate our total Opex for the field life as a minimum of 45.83 MMbbl (275 bcf) × 8 equaling 366.64 and maximum of 45.83 MMbbl × 15 equaling 687.45 MM$.

This total cost should be prorated proportionally to production for the remaining years.

For the peak rates on plateau, our Opex yearly would be 50% × 366.64/5 equaling 36.664 MM$ or 50% × 687.45/5 equaling 68.745 MM$ and prorated for the remaining years. Again, the Opex distribution during the years is merely indicative and should be based on real operating costs for current and previous projects.

12.8.7 Project Timing and Input to Economic Analysis

A SPAR will require 4–5 years from exploration to production on average. We can drill the wells in 2012 & 2013. Our project may start up production in 2015 assuming we have already drilled our exploratory well in 2010.

Date	Oil Production rate	Gas Production rate	Exploration Drilling	Appraisal Drilling	Development Drilling	Capex	Capex	Capex	Abandonment	Opex	Comments
	stb/day	MMscf/day	MM$	MMS	MM$	MM$	MM$	MM$	MM$	MM$	
			min-max	min-max	min-max	Facility	Subsea	pipeline			
2010			35-64								1 Exploratory
2011											No appraisals well
2012					(63-115) × 1	200	136	8.8			1 well
2013					(99-115) × 1	200		28.98			1 well
2014							68	512			
2015	1.20	120.00								18.33-34.37	
2016	2.40	240.00								36.66.68.74	
2017	2.40	240.00								36.66.68.74	
2018	2.40	240.00								36.66.68.74	
2019	2.40	240.00								36.66.68.74	
2020	2.00	200.00								36.66.68.74	
2021	1.50	150.00								18.33-34.37	
2022	1.02	102.00								18.33-34.37	
2023	0.60	60.00								18.33-34.37	
2024	0.30	30.00								9.17-17.19	
2025	0.12	12.00								9.17-17.19	
2026	0.06	6.00								9.17-17.19	
2027	0.03	3.00								9.17-17.19	
2028	0.02	2.00								9.17-17.19	
2029	0.02	1.00								9.17-17.19	
2030										9.17-17.19	
2031											
2032											
Total			35-64		126-230	400	68	549.78			

Figure 12.58 Economic analysis inputs.

12.9 CASE 9—ACCUMULATION WITH OIL AND GAS RESERVOIR COMBINED

Make an economical analysis for a prospect with a volume in place of 500 MMbbl of oil, low dissolved gas, and 500 bcf of dry gas, situated in the deepwater sector of Gulf of Mexico with a water depth of 5500 ft. It is Pliocene and Middle Miocene age with a total depth of 25,000 ft and an area of 3420 acres. Net pay equals 300 ft for the oil and 100 ft for the gas. No exploratory wells, and all data is provided from seismic.

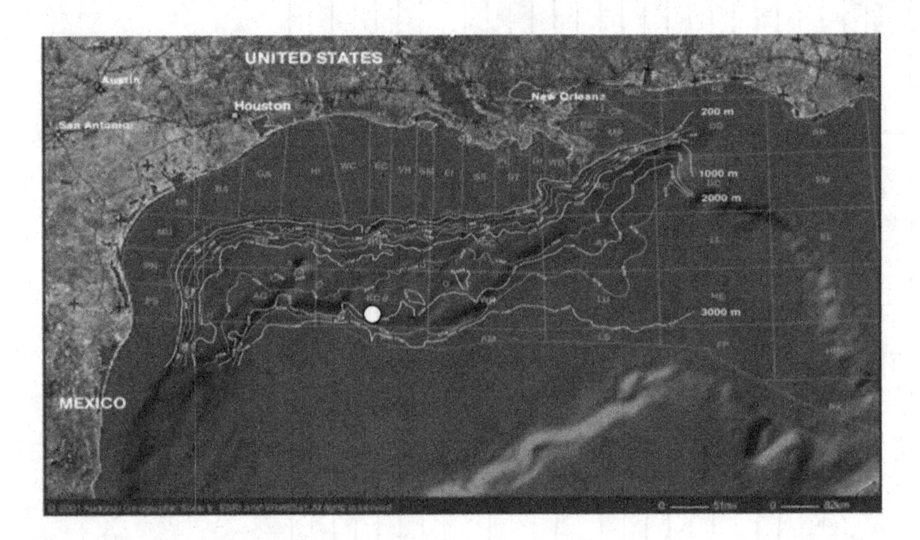

Figure 12.59 Field location map.

12.9.1 Recovery Factor Estimation

The first thing will be to estimate the recover factor for this accumulation in terms of oil and gas. In the example shown in the recovery factor chapter, it is possible to see a trend of recovery factor by depth and by play. We can see that we should expect recoveries of 75 % for gas in the Pliocene and 30% for oil in the middle Miocene play with reservoirs at a depth of 25,000 ft. Therefore, our recoverable volume will be 65% × 500 bcf equaling 325 bcf and 35% × 500 MMbbl equaling 175 MMbbl.

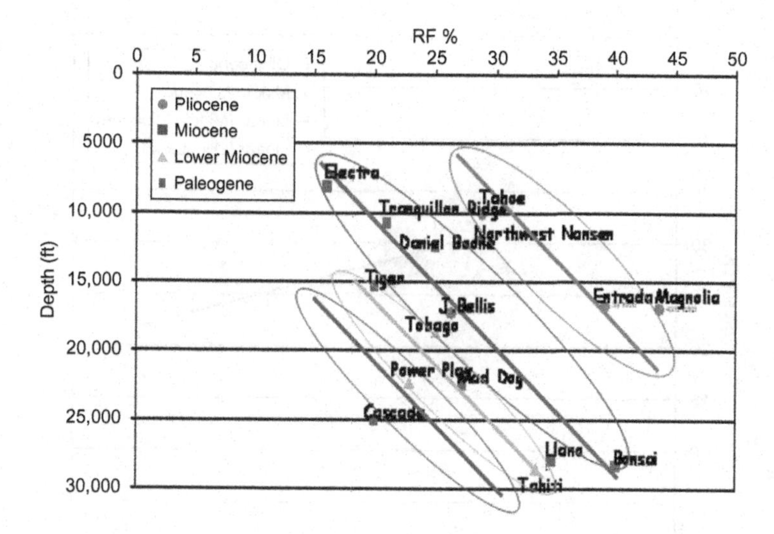

Figure 12.60 Example of a trend deriving using analogs and grouping parameters.

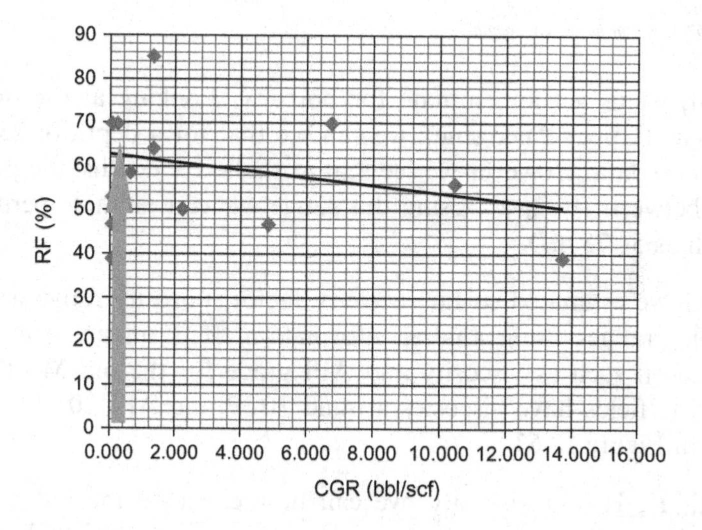

Figure 12.61 RF for a dry gas (CGR = 0).

12.9.2 Rock/Fluid Properties and Well Count Estimation

Next step is to derive the number of wells and their initial rate. To do this, we first need to estimate some rock and fluid properties.

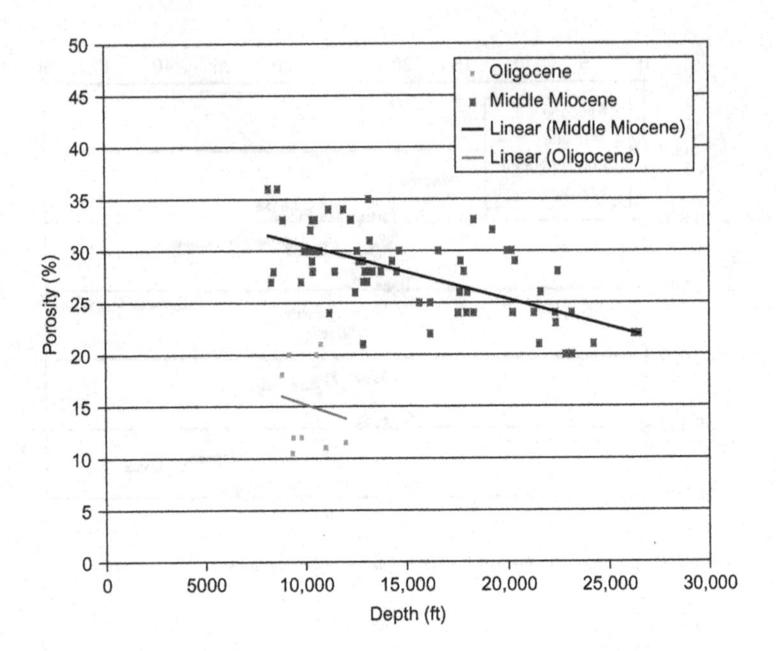

Figure 12.62 Trend for porosity vs. depth.

First, let us get an estimate for porosity. Looking at the dots for Oligocene (Upper Paleogene), we can see that for a depth of 25,000 ft there is no data. However we see that at shallower depths, the porosity varies between 10–20%. Using this range, we can estimate permeability such as 5–75 mD.

We have estimated so far recovery factor, porosity range and permeability ranges using analog information. It is missing some fluid information such as viscosity and API grade for the oil. Most of the Paleogene fields have viscosity around 10 cP and API 20–22 as suggested in Figure 12.63.

With K, H and viscosity, we can now calculate the index kh/mu and estimate the recovery per well and therefore the total number of producer wells. K = 5–75 mD, H = 300 ft and Viscosity = 10 cP. Kh/mu (min) = 5 × 300/10 = 150 mD ft/cP and kh/mu (max) = 75 × 300/10 = 2250 mD ft/cP.

The numbers 150 and 2250 would lie in the lower range of the plot in Figure 12.62, with extremely low recovery per well, as expected due mainly to the viscosity of the oil (i.e. heavy, 22 API, 10 cP) and low

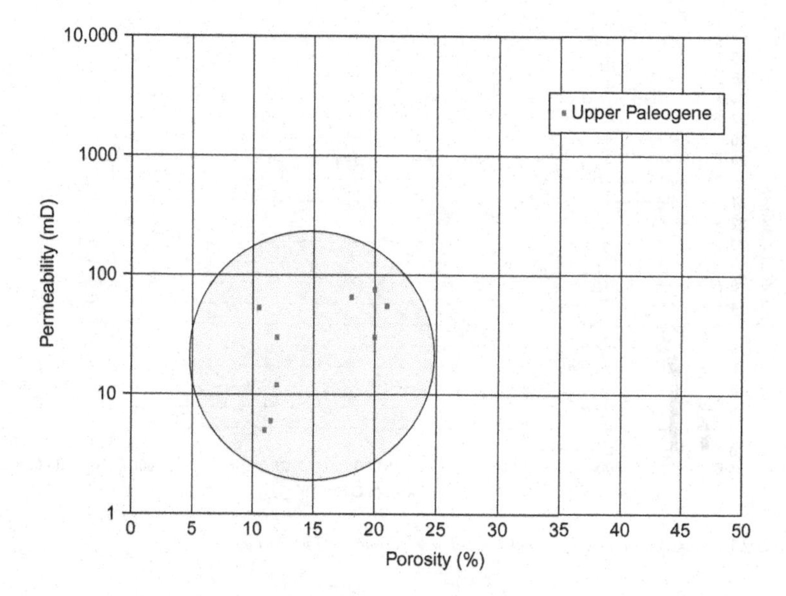

Figure 12.63 Permeability for Oligocene play from a database with analog information.

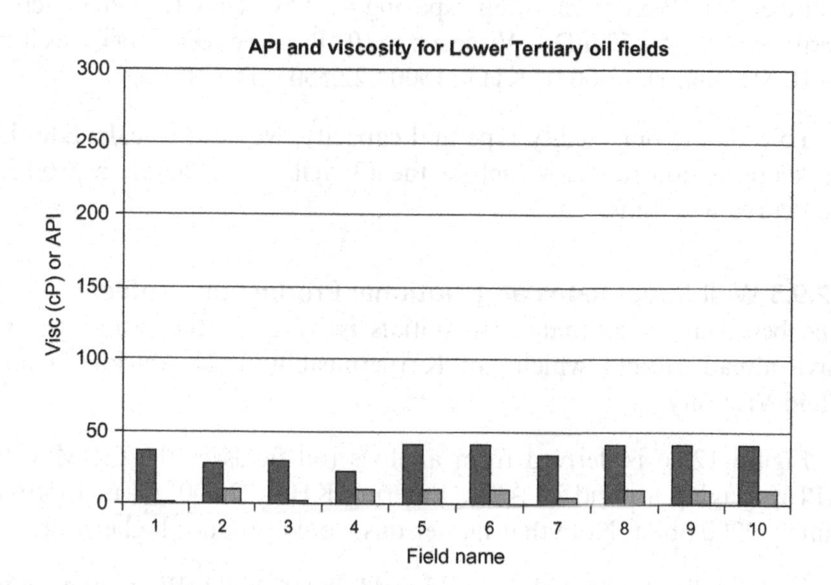

Figure 12.64 Viscosity for fields in the Lower Tertiary.

permeability reservoir. The UR/well would then be between 2 and 10 MMbbl of oil leading to a low case of 125/2, or 63 wells and a high case with 13 wells. The low case with 63 wells would be uneconomical. Therefore, let's go ahead with the high case with 13 wells.

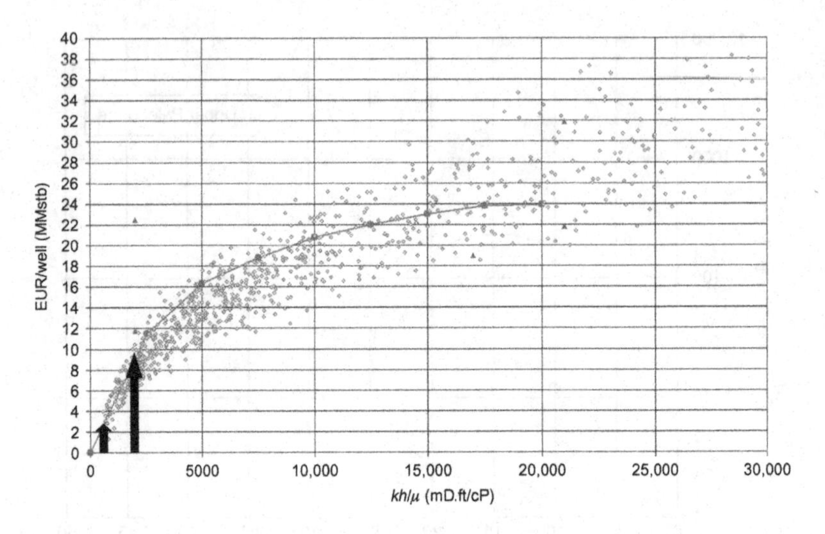

Figure 12.65 Estimation of number of wells based on rock and fluid properties.

So far, we have estimated the following parameters for this project: Number of Wells—13, well spacing = 3420 acres/13 = 263 acres, Permeability—5–75 mD, Viscosity = 10 cP, Recovery per well = 2–10 MMbbl, H = 300 ft, KH = 1500–22,500 mD × ft.

To estimate our facility type and capacity, we need to calculate the peak production rates for each of the 13 wells and generate a production curve with time.

12.9.3 Well Initial Rates and Notional Production Profile

The best way to estimate well initials is by using the properties we have already found which are K (permeability), H (thickness) and Fluid Viscosity.

Figure 12.66 is derived from analogs (oil fields in the GOM with API averaging around 24 API). Entering KH = 22,500, Qi is approximately 2200 bbl/d. Note that lighter oils would produce higher rates.

So 13 wells at 2200 bbl/d would yield 28,600 bbl/d. We now need to create a production curve that produces a plateau of 28,600 bbl/d and total reserves of 125 MMbbl.

Figure 12.67 would fit as our production scenario. In the first year, the production is lower because not all wells will be on production. (In this case, only half of the total, or about 6 wells.)

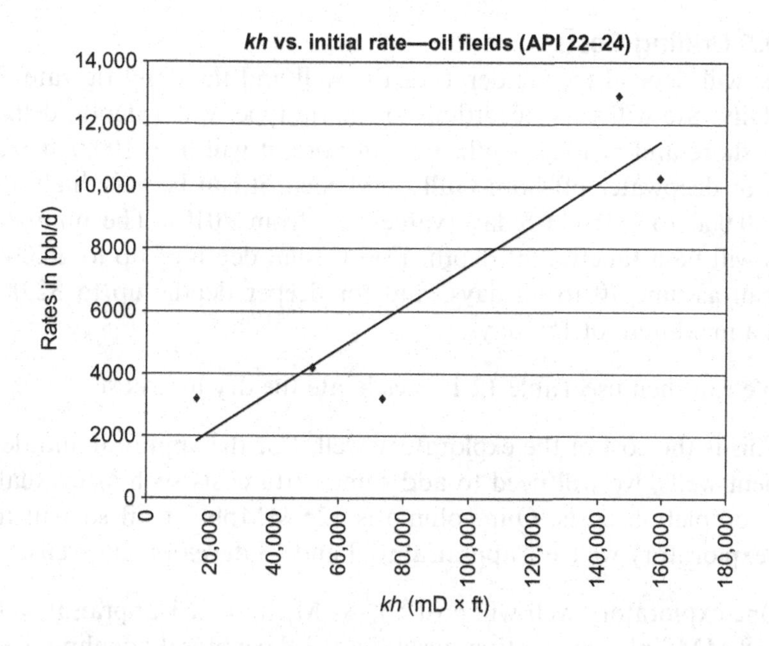

Figure 12.66 Estimation of well rates based on rock and fluid properties.

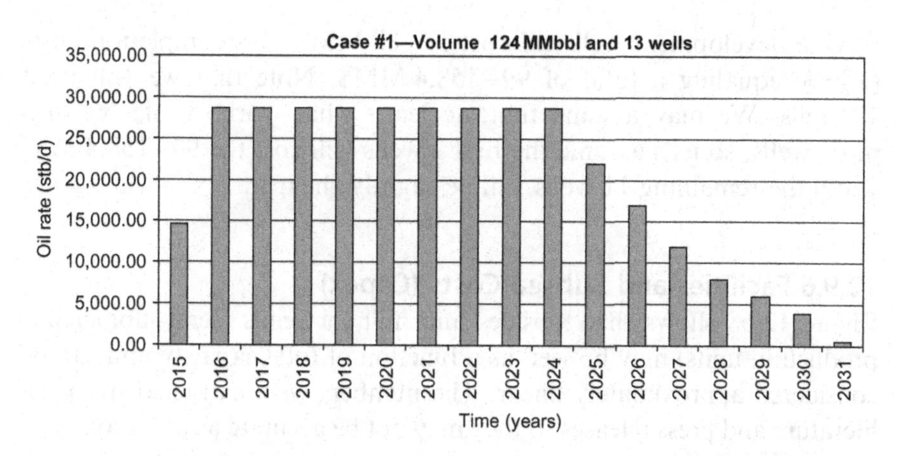

Figure 12.67 Production forecasting.

12.9.4 Costs: Drillex, Capex and Opex

Now it is necessary to estimate the cost to drill the wells, build the facilities, and evacuate the production as well as the operating costs.

12.9.5 Drilling Costs

Costs will depend on the depth of the well and the daily rig rate. The rig daily rate will vary according to the rig type, water depth, distance from shore and drilling depth. For onshore, it will be <100,000 $/day, and for deepwater offshore Gulf of Mexico, it can be very high at up to 600,000 to 800,000 $/day (values are from 2010). The number of days will be a function of depth. For a usual depth of up to 20,000 ft, we can assume 70 to 80 days, and for deeper depths up to 32,000 ft with a maximum of 150 days.

We can then use Table 12.1 to estimate the dry hole cost.

This is the cost of the exploratory well. For the appraisal and development wells, we will need to add some extra costs such as evaluation and completion costs. Our volume is 125 MMbbl of oil so will need one exploratory well + 1 appraisal well and 13 development wells.

One exploratory well will cost 55–88 MM$, and 1 appraisal well at 55–88 MM$ plus evaluation costs (i.e. 15 days more) equaling a total of 62.5–100 MM$.

One development well will cost 55–88 MM$ plus completion costs (+80%) equaling a total of 99–158.4 MM$. Note that we will need 13 wells. We may assume that we learn what works while we drill more wells, so let's assume the first 2 wells will cost the 99–158.4 MM $ and the remaining 11 wells will be slightly cheaper.

12.9.6 Facilities and Subsea Costs (Capex)

Figure 12.68 shows that topsides and hull for semis (semi-submersible production units) may be seen as a function of total capacity and can be considered approximately linear. The numbers were extracted from the literature and press releases so they may not be accurate as of today.

The x-axis represents unit capacity in thousands of barrels per day and the y-axis is the cost (fabrication + installation) in MM$. We want our unit to be sized to accommodate 30 kbbl/d. Using the equations as shown in Figure 12.68, we get 30 kbbl/d, hull cost is approximately 400 MM$ and the top side is about 552 MM$, making a total of 958 MM$. We may want to add some contingency costs here around at about 25% or 1278 MM$.

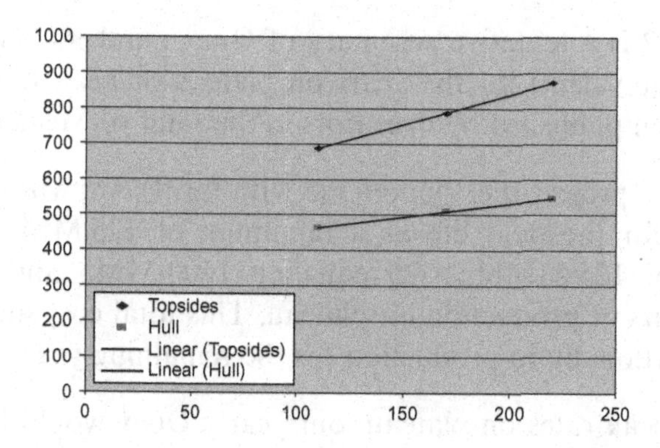

Figure 12.68 Hull and topsides Capex estimation in MM\$ versus FPU capacity in kbbl/day.

The next step is to calculate the subsea costs and export pipelines costs. For 13 wells, we will need two manifolds, one with 6 wells and another one for 7 wells, plus inter field flow lines and risers. These costs usually come from a previous project and/or vendor. Let's assume some factors here to estimate these costs:

- Well templates and manifolds = 34 MM\$/well
- SCR Production Riser = 2 risers per manifold = 4 Risers at 0.35 MM\$/(″ × 1000 ft)
- SCR Export Riser = 1 at 0.28 MM\$/(″ × 1000 ft)
- Flow lines and export pipelines at 0.23 and 0.16 MM\$/ (″ × miles). Let's assume flow lines length = 5 miles and export pipeline length = 30 miles

These are estimated factors as provided by vendors and publications. Templates and manifolds will equal 34 × 13 wells or 442 MM\$. Production Risers will equal 4 × 8 × water depth (i.e. 1.5 × 5500 ft) × 0.35/1000 × 10″ equaling 924 MM\$. The export riser is 0.28 × 5500 ft/1000 ft × 10″ equaling 15.4 MM\$, and FL plus export pipelines equals 0.23 × 5 miles + 0.16 × 30 miles × 8″ or 48 MM\$.

12.9.7 Operating Costs (Opex)

The operating costs are more complex and difficult to estimate using only charts or factors. They are usually a function of the peak production rate and tariff rates paid to export production via third party pipelines.

Table 12.2 is a tentative summary of Opex numbers per BOE (barrels of oil equivalent) for the years on plateau or peak rates based on what has been published by operators in the Gulf of Mexico (GOM).

So for this project (i.e. standalone with tariff), we will estimate our total Opex for the field life as a minimum of 125 MMbbl × 12 and maximum of 125 MMbbl × 25 equaling 1500 MM$ and 3125 MM$ for the 9 years of production on plateau. This total cost should be prorated proportionally to production for the remaining years.

For the peak rates on plateau, our yearly Opex would be 1500/9 or 3125/9 and prorated for the remaining years. A minimum of 25% Opex is recommended for the years where the production is lower than 25% of peak to take into account the maintenance costs and other costs that are not a direct function of the production.

12.9.8 Project Timing and Input to Economic Analysis

We have now calculated most of the input we need for our economic analysis and to the end of our VV (volume to value) process (Table 12.3).

The timing of the project is just as crucial as all the other inputs, but it is also flexible depending how aggressively the company wants to drill the wells and invest in advance of the fabrication of facilities.

Ideally, we will drill at least one well per year and consider a period of time to fabricate the facilities and have them in place.

The table provides an overview for the total project time from exploration to first production. In the best case scenario, this time won't be less than 5 years. For a big project, it can reach 10 or more years from the exploratory well to first production.

Now it is possible to place all the information we need to calculate net present value (NPV) for the opportunity and other economic parameters. Figure 12.69 summarizes the inputs that any economic package will require, basically the revenues (volumes) and the expenditures. Note that in the Figure 12.69, there are min and max numbers for the costs. The user may choose to pick an average number before running the economics.

Date	Oil Production rate	Gas Production rate	Exploration Drilling	Appraisal Drilling	Development Drilling	Capex	Capex	Capex	Abandonment	Opex	Comments
	stb/day	MMscf/day	MM$	MM$	MM$	MM$	MM$	MM$	MM$	MM$	
			min-max	min-max	min-max	Facility	Subsea	pipeline			
2010			55-88								1 Exploratory
2011				62.5-100	(99-158) × 1	255.6					1 Appraisal+1 well
2012					(99-158) × 3	255.6					3 wells
2013					(99-158) × 3	255.6	147.3333	924			3 wells
2014					(99-158) × 3	255.6	147.3333	15.4			3 wells
2015	14,300.00				(99-158) × 3	255.6	147.3333	48			3 wells
2016	28,600.00									83.33-173.61	
2017	28,600.00									166.66-347.22	
2018	28,600.00									166.66-347.22	
2019	28,600.00									166.66-347.22	
2020	28,600.00									166.66-347.22	
2021	28,600.00									166.66-347.22	
2022	28,600.00									166.66-347.22	
2023	28,600.00									166.66-347.22	
2024	28,600.00									166.66-347.22	
2025	22,600.00									83.33-173.61	
2026	17,000.00									41.67-86.81	
2027	12,000.00									41.67-86.81	
2028	8,000.00									41.67-86.81	
2029	6,000.00									41.67-86.81	
2030	4,000.00									41.67-86.81	
2031	1,000.00								127.60	41.67-86.81	10% of Total Facility Capex
Total	volume = 125 MMbbl		55-88	62.5-100	1287-2054	1276	442				

Figure 12.69 Economic analysis inputs.

12.10 CASE 10—EXAMPLE OF FIELDS EVACUATED TO A SHARED FACILITY

Make an economical analysis for an oil prospect, with a volume in place of 1000 MMbbl of oil, low dissolved gas, situated in a remote region in deepwater Gulf of Mexico with a water depth of 5500 ft.

It is of Lower Tertiary (Paleogene) age and situated at a total depth of 25,000 ft and area of 6800 acres. Net pay is approximately 300 ft with no exploratory wells, and all data is provided from seismic.

For this case all the assumptions calculated in the case study 2 will apply in terms of RF, EUR per well, well initials, and rock and fluid properties.

The production will be evacuated via a facility that will be shared with another party in terms of cost and production.

The difference will be how to model the shared facility. Capex and Opex will be spent proportionally to the volumes or production going to the facility from each of the companies.

The facility will have to be sized to a bigger capacity to accommodate production from two fields.

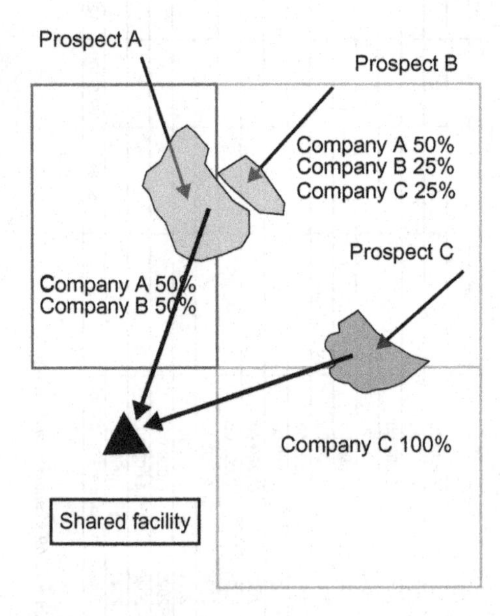

Figure 12.70 Example of a shared facility for Prospect A & C.

12.10.1 Facilities and Subsea Costs (Capex)

Figure 12.71 will be used again to provide an estimation of Capex.

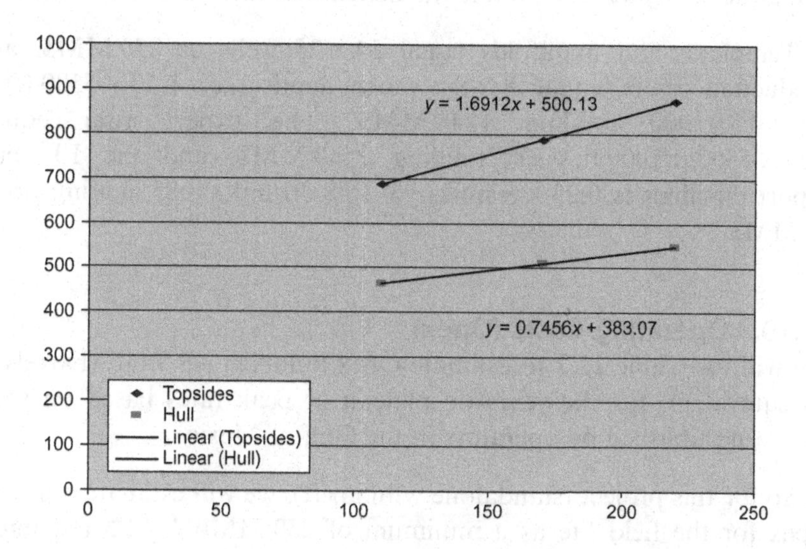

Figure 12.71 Hull and topsides Capex estimation MM$ versus FPU capacity.

We want our unit to be sized to accommodate 60 kbbl/d. Using the equations as shown in Figure 12.71, we get 60 kbbl/d, hull cost is about 450 MM$ and top side is about 650 MM$ with a total equaling 1100 MM$. We may want to add some contingency costs here around of 25% or 1375 MM$.

Note that although the rates increased twofold the Capex costs did not double.

Next step is to calculate the subsea costs and export pipelines costs. For 25 wells, we will need four manifolds (three with 6 wells and another one for 7 wells) plus inter field flow lines and risers. Let's assume the some factors used in case study 1 here to estimate these costs. Costs should look like:

- Well templates and manifolds = 34 MM$/well
- SCR Production Riser = 2 risers per manifold = 8 Risers at 0.35 MM$/(″ × 1000 ft)
- SCR Export Riser = 1 at 0.28 MM$/(″ × 1000 ft)—assuming that the riser has enough capacity to handle the increase in production.

- Flow lines and Export Pipelines at 0.23 and 0.16 MM$/ ($'' \times$ miles). Let's assume flow lines length = 5 miles and export pipeline length = 30 miles (let's keep to the same as for case study 1 assuming the flow lines can handle the increase of rate up to 60 kbbl/d).

Templates and manifolds equal 34×25 wells or 850 MM$, and production risers equal $8 \times 8 \times$ water depth (i.e. 1.50×5500 ft) $\times$ $10'' \times 0.35/1000$ totaling 1848 MM$. The export riser equals 0.28×5500 ft/1000 ft $\times 10''$ totaling 1540 MM$, and the FL plus export pipelines is 0.23×5 miles $+ 0.16 \times 30$ miles $\times 8''$, adding up to 48 MM$.

12.10.2 Operating Costs (Opex)

We will use Table 12.2 to estimate Opex numbers per BOE (barrels of oil equivalent) for the years on plateau or peak rates based on what has been published by operators in the Gulf of Mexico.

So for this project (standalone with tariff), we will estimate our total Opex for the field life as a minimum of 250 MMbbl $\times$ 12 and maximum of 250 MMbbl $\times$ 25 equaling 3000 MM$ and 6250 MM$ for the 9 years of production on plateau (until 2014). This total cost should be prorated proportionally to production for the remaining years.

For the peak rates on plateau, our yearly Opex would be 3000/9 or 6250/9 and prorated for the remaining years. A minimum of 25% Opex is recommended for the years where the production is lower than 25% of peak to take into account the maintenance costs and other costs that are not a direct function of the production.

12.10.3 Project Timing and Input to Economic Analysis

We have now calculated most of the input we need for our economic analysis and to the end of our VV (volume to value) process. The timing of the project (first production) will be longer than 1 year than in case study 1 because we will need an extra appraisal well as the volume is now two times bigger. This second appraisal well may be drilled in the second or third year pushing all the development wells one year later and thus delaying the start up of first production. We will also assume that the extra development wells will be drilled simultaneously by a second rig.

Date	Oil Production rate	Gas Production rate	Exploration Drilling	Appraisal Drilling	Development Drilling	Capex	Capex	Capex	Abandonment	Opex	Comments
	Stb/day	MMScf/day	MM$	MMS	MM$	MM$	MM$	MM$	MM$	MM$	
			min-max	min-max	min-max	Facility	Subsea	pipeline			
2010			55-88								1 Exploratory
2011				62.5-100		275					2 Appraisal+1 well
2012				62.5-100	(99-158) × 1	275					6 wells (2 rigs)
2013					(99-158) × 6	275	283.33	1848			6 wells
2014					(99-158) × 6	275	288.33	15.5			6 wells
2015					(99-158) × 6	275	288.33	48			6 wells
2016	30,000.00				(99-158) × 6					166.50	
2017	60,000.00									333.00	Opex max with Tariff
2018	60,000.00									333.00	
2019	60,000.00									333.00	
2020	60,000.00									333.00	
2021	60,000.00									333.00	
2022	60,000.00									333.00	
2023	60,000.00									333.00	
2024	60,000.00									333.00	
2025	60,000.00									333.00	
2026	48,500.00									333.00	
2027	37,000.00									166.50	
2028	27,000.00									166.50	
2029	19,000.00									166.50	
2030	12,000.00									83.25	
2031	7,500.00									83.25	
2032	3,700.00								137.50	83.25	10% of Total Facility Capex
Total			55-88	125-200	2475-3950	1375	850	1911.5			

Figure 12.72 Economic analysis inputs.

12.11 CASE 11—EXAMPLE OF A TIGHT GAS OFFSHORE RESERVOIR

Make an economical analysis for an oil prospect, with a volume in place of 300 MMbbl of oil, low dissolved gas, situated in deepwater Gulf of Mexico with a water depth of 2500 ft. It is Lower Miocene age and total depth of 25,000 ft with an area of 1200 acres. Net pay equals 100 ft with no exploratory wells, and all data is provided from seismic. The prospect is just 15 miles of the existing Marco Polo Platform.

This example is a good case for a subsea tieback due to the small

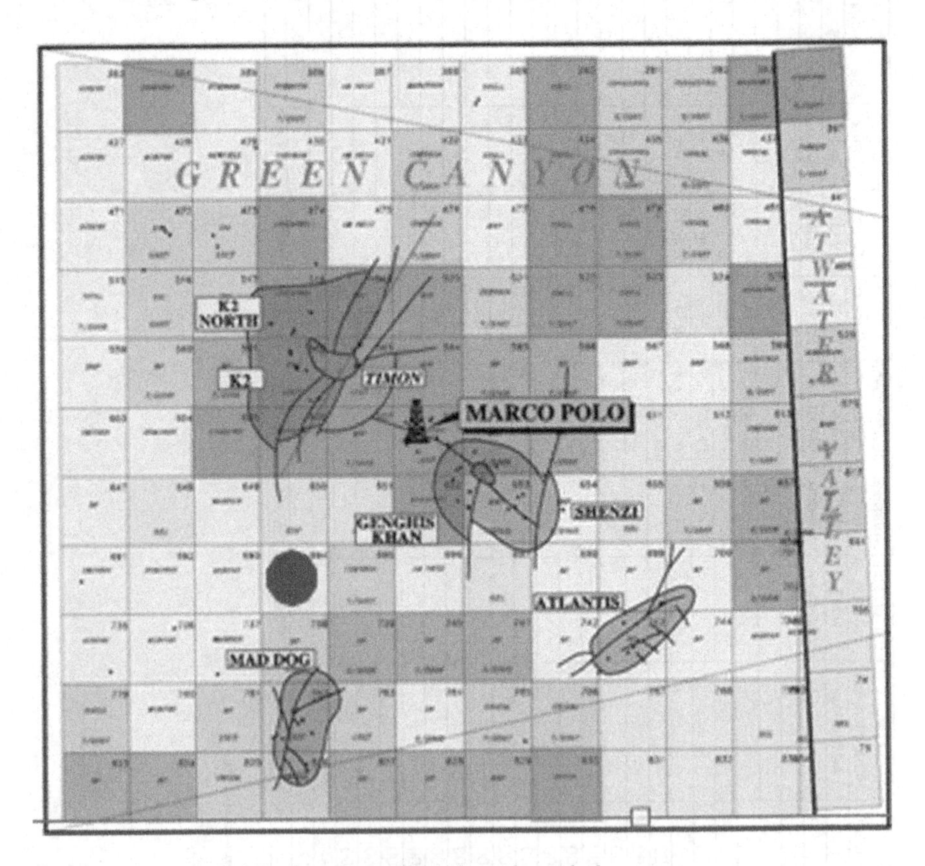

Figure 12.73 Location of a light oil prospect, candidate for a subsea tieback.

volume and also the proximity of an existing infrastructure.

12.11.1 Recovery Factor Estimation

The first thing will be to estimate the recovery factor for this light oil accumulation. We can see we should expect recoveries of 30% for a Miocene play with reservoirs at depth of 25,000 ft. Therefore our recoverable volume will be 30% × 300 equaling 90 MMbbl.

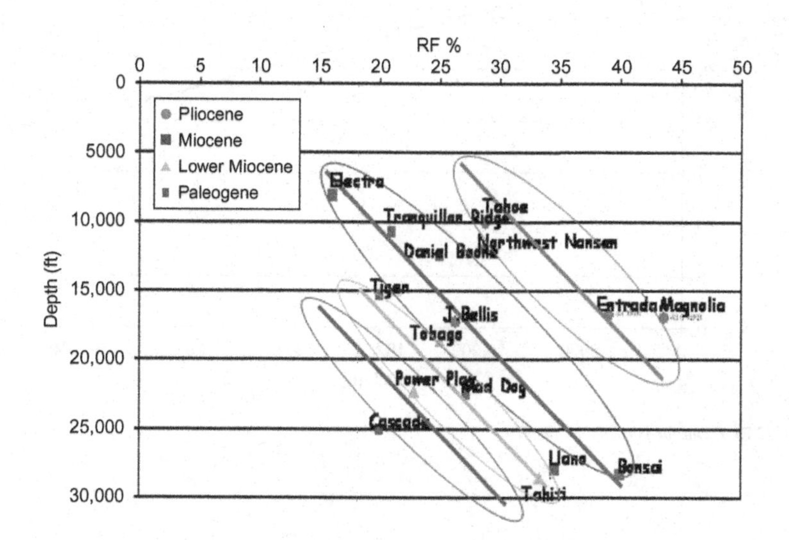

Figure 12.74 Example of a trend deriving using analogs and grouping parameters.

In this case, it would be prudent to look at ranges of expected recovery from the nearby fields such as Mad Dog and Genghis Khan. Commercial databases and searching literature suggests Mad Dog is at 25–32% RF and Genghis Khan is at 16.2%. Although our 30% estimation is not bad, we could also run a lower case for a 16% recovery factor making 16% × 300, equaling 48 MMbbl.

12.11.2 Rock/Fluid Properties and Well Count Estimation

Next step is to derive the number of wells and their initial rate. To do this, we first need to estimate some rock and fluid properties.

First, let us get an estimate for porosity. Looking at the dots for Miocene we can see that for a depth of 25,000 ft, porosity is around 22% and permeability from 100 to 500 mD.

We have estimated so far recovery factor, porosity range and permeability ranges using analog information. We are missing some fluid

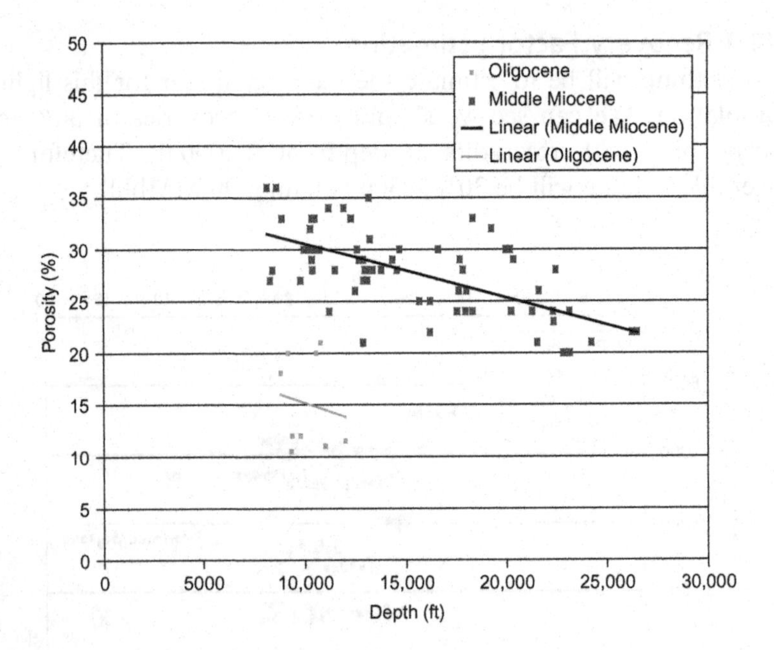

Figure 12.75 Trend for porosity vs. depth.

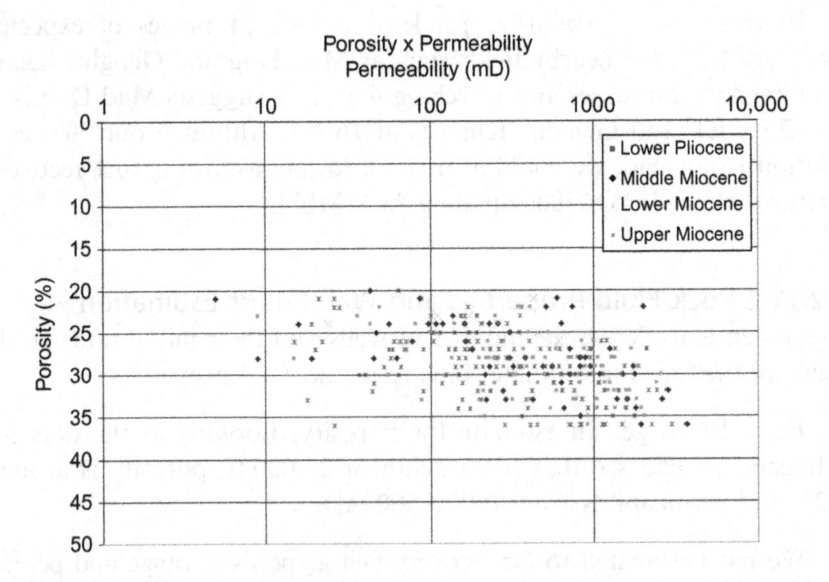

Figure 12.76 Porosity versus permeability trend.

information such as viscosity and API grade for the oil. Again, let us use Mad Dog fluid as an analog. Mad Dog has API 32 and a viscosity of 2 cP.

With K, H and Viscosity, we can now calculate the index kh/mu and estimate the recovery per well and therefore the total number of producer wells. K = 100–500 mD, H = 100 ft and Viscosity = 2 cP. Kh/mu (min) equals 100 × 300/2 or 15,000 mD ft/cP and kh/mu (max) equals 500 × 100/2 or 25,000 mD ft/cP.

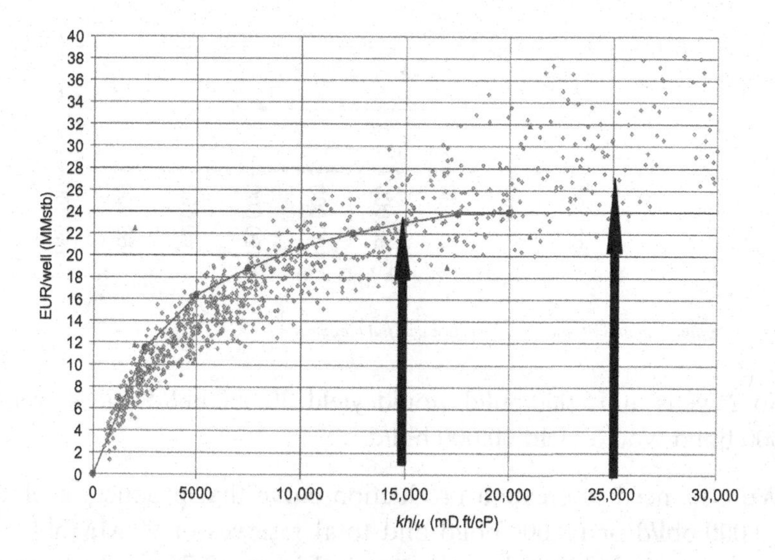

Figure 12.77 Estimation of number of wells based on rock and fluid properties.

The UR/well would then be between 23 and 28 MMbbl of oil (i.e. average 25 MMbbl) leading to a low case with a low volume of 48/25 or 2 wells and a high case with 90/25 or 4 wells.

12.11.3 Well Initial Rates and Notional Production Profile

The best way to estimate well initials is using the properties we have already found: K (permeability), H (thickness) and Fluid Viscosity. Figure 12.78 is derived from analogs (i.e. oil fields in the GOM with API averaging around 24 API). Our fluid will have a higher API, 32, and a lower viscosity, as we are being conservative.

Entering KH = 15,000, Qi is about 10,000 bbl/d, and KH = 25,000 rates and could be up to 20,000 bbl/d. Let's use the average 15,000 bbl/d.

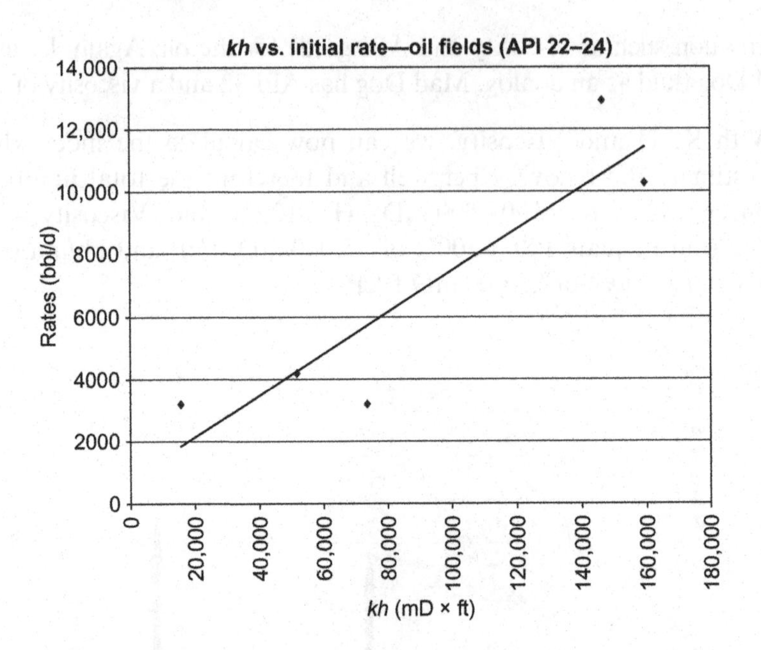

Figure 12.78 Estimation of well rates based on rock and fluid properties.

So 2 wells at 15,000 bbl/d would yield 30,000 bbl/d and 4 wells at 15,000 bbl/d would yield 60,000 bbl/d.

We now need to create a production curve that produces a plateau of 30,000 bbl/d or 60,000 bbl/d and total reserves of 90 MMbbl for a high case or 48 MMbbl for a low case. Figure 12.79 would fit as our production scenario for 60,000 bbl/d with 4 wells.

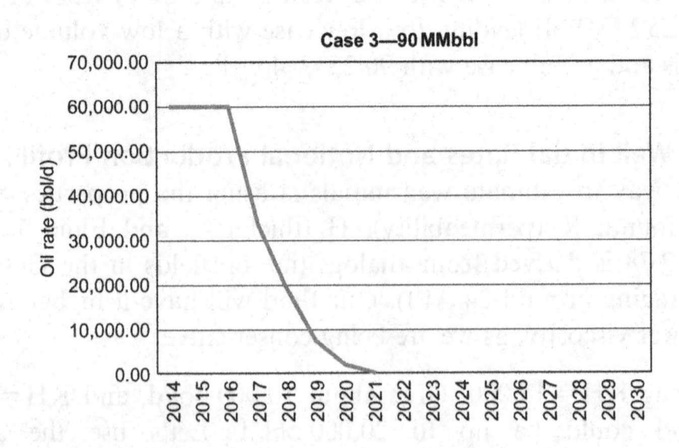

Figure 12.79 Production forecasting.

12.11.4 Costs: Drillex, Capex and Opex

Now it is necessary to estimate the cost to drill the 2 wells and evacuate the production as well as the operating costs.

12.11.5 Drilling Costs

Use Table 12.1 to estimate the dry hole cost.

For a depth of 25,000 ft, the drilling costs will be 55–88 MM$ for the exploratory well. As the volumes are lower than 100 MMbbl, we will not need an appraisal well. One development well will cost 55–88 MM$ plus completion costs (+ 80%) totaling 99–158.4 MM$. Note that we will need 2 wells (or 4 in the high case scenario).

12.11.6 Facilities and Subsea Costs (Capex)

In this example, we will have only subsea and pipeline costs as there is no hub or facility to be built because we will tie back production to the Marco Polo Platform. Next step is to calculate the subsea costs and export pipelines costs. For 2 or 4 wells we will need only one manifold plus export pipeline to the Marco Polo Platform and the riser. Costs should look like:

- Well templates and manifolds = 34 MM$/well
- SCR Production Riser = at 0.35 MM$/(″ × 1000 ft)
- Export Pipelines at 0.16 MM$/ (″ × miles). Let's assume that the export pipeline length = 15 miles

Templates and manifolds = 34 × 4 wells equaling 136 MM$ and production risers are 1 × 1 × water depth (i.e. 1.5 × 5500 ft) × 10″ × 0.35/1000 totaling 28.8 MM$. Export pipeline is 0.16 × 15 miles × 8″ equaling 19.2 MM$.

12.11.7 Operating Costs (Opex)

We will use the value for tieback provided by Table 12.2 with tariff 8–13 $/boe.

So for this project (subsea tieback with tariff), we will estimate our total Opex for the field life as a minimum of 90 MMbbl × 8 and maximum of 90 MMbbl × 13 or 720 MM$ and 1170 MM$ for the 3 years of production on plateau.

This total cost should be prorated proportionally to production for the remaining years. For the peak rates on plateau, our yearly Opex would be 75% × 720/3 equaling 180 or 75% × 1170/3 equaling 292.5 and prorated for the remaining years.

Again, a minimum of 25% Opex is recommended for the years where the production is lower than 25% of peak rate.

In the two previous cases, case study 1 and case study 2 at the beginning of this chapter, we concentrated the Opex on the plateau years. This is not totally correct. Although most of the Opex will be spent in the plateau years, some will be spent in the remaining years.

12.11.8 Project Timing and Input to Economic Analysis

A subsea tieback will require 3 years from exploration to production on average. We can drill all four wells during this time (i.e. 2 wells per year). Our project may start up production in 2014 assuming we drill our exploratory well in 2011 (Table 12.3).

Date	Oil Production rate	Gas Production rate	Exploration Drilling	Appraisal Drilling	Development Drilling	Capex	Capex	Capex	Abandonment	Opex	Comments
	stb/day	MMscf/day	MM$	MMS	MM$	MM$	MM$	MM$	MM$	MM$	
			min-max	min-max	min-max	Facility	Subsea	pipeline			
2010			55-88								1 Exploratory
2011											No appraisals
2012					(99-158) × 2		136	19.2			2 wells
2013					(99-158) × 2			28.8			2 wells
2014	28,000.00									90-146	Opex max with Tariff
2015	28,000.00									90-146	
2016	28,000.00									90-146	
2017	28,000.00									90-146	
2018	28,000.00									90-146	
2019	28,000.00									90-146	
2020	27,000.00									90-146	
2021	20,000.00									45-73	
2022	13,500.00									45-73	
2023	8,000.00									45-73	
2024	5,000.00									22-5-36.6	
2025	3,000.00									22-5-36.6	
2026	1,500.00									22-5-36.6	
2027	800.00									22-5-36.6	
2028											
2029											
2030											
2031											
2032											
Total			55-88		396-632		134	48			

Figure 12.80 Economic analysis inputs.

12.12 CASE 12—EXAMPLE OF A VERY LARGE OFFSHORE FIELD

Make an economical analysis for an oil prospect, with a volume in place of 1000 MMbbl of oil, low dissolved gas, situated in deepwater Gulf of Mexico with a water depth of 5500 ft. It is Lower Miocene age and total depth of 15,000 ft with an area of 1200 acres. Net pay equals 200 ft, and no exploratory wells exist. All data is provided from seismic.

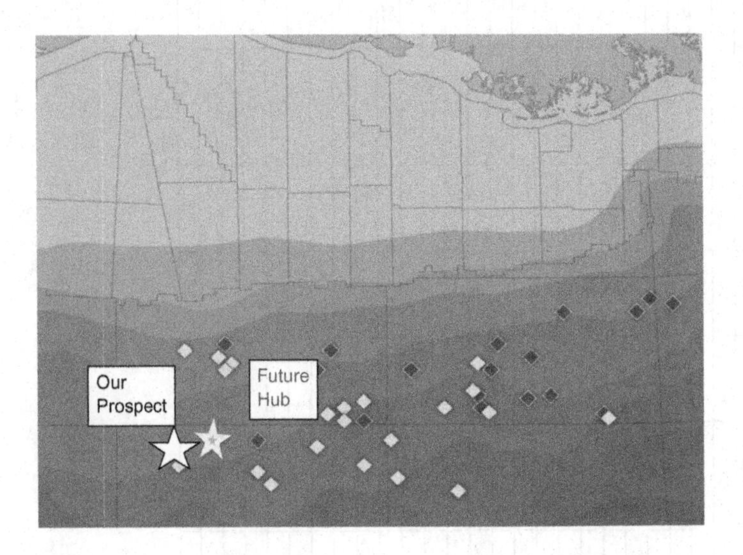

Figure 12.81 Location of a light oil prospect in deepwater GOM.

The diamonds represent potential prospects in Miocene (in gray color) and Lower Tertiary (in white).

12.12.1 Recovery Factor Estimation

The first thing will be to estimate the recover factor for this light oil accumulation. We should expect recoveries of 19% for a lower Miocene play with reservoirs at depth of 15,000 ft. Therefore, our recoverable volume will be 19% × 1000 or 190 MMbbl.

12.12.2 Rock/Fluid Properties and Well Count Estimation

Next step is to derive the number of wells and their initial rate. To do this, we first need to estimate some rock and fluid properties.

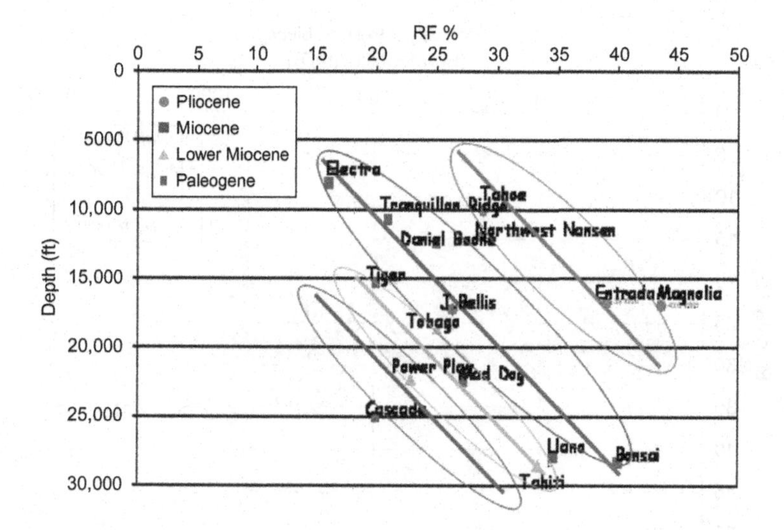

Figure 12.82 Example of a trend deriving using analogs and grouping parameters.

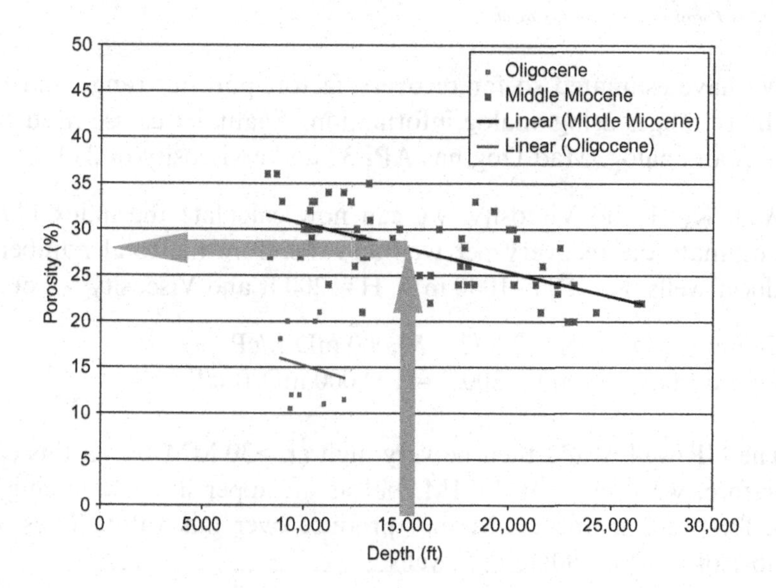

Figure 12.83 Trend for porosity vs. depth.

First, let's get an estimate for porosity. Looking at the dots for Miocene, we can see that for a depth of 15,000 ft, porosity is around 28% and permeability from 500 to 1000 mD.

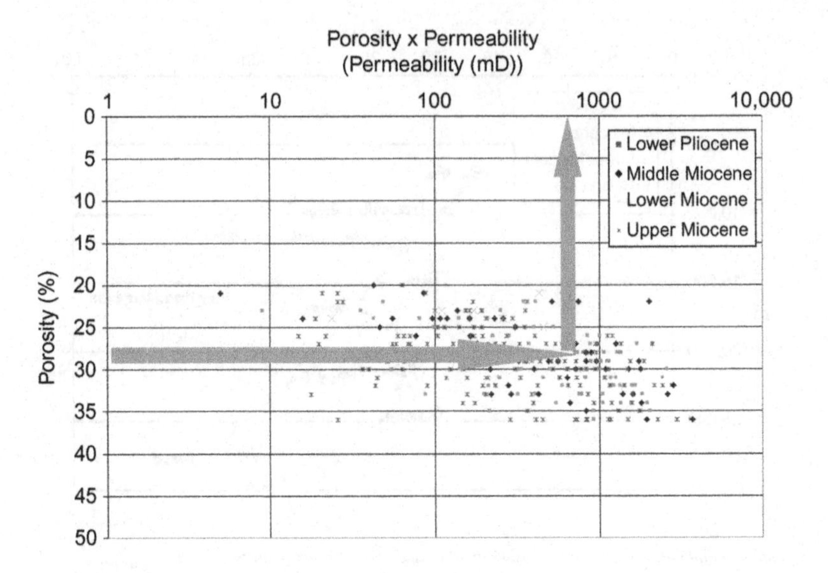

Figure 12.84 Porosity vs permeability trend.

We have estimated so far recovery factor, porosity range and permeability ranges using analog information. Again, let us use Mad Dog fluid as an analog. Mad Dog has API 32 and a viscosity of 2 cP.

With K, H and Viscosity, we can now calculate the index kh/mu and estimate the recovery per well and therefore the total number of producer wells. K = 500–1000 mD, H = 200 ft and Viscosity = 2 cP.

- Kh/mu (min) = 500 × 200/2 = 50,000 mD ft/cP
- Kh/mu (max) = 1000 × 200/2 = 100,000 mD ft/cP

The UR/well would then be very high at >30 MM bbl in this case. Therefore, we keep it at 30MM bbl as an upper limit, assuming we have few wells in the GOM that produce over this value. Thus, our number of wells is 190/30 or 7 wells.

12.12.3 Well Initial Rates and Notional Production Profile

The best way to estimate well initials is using the properties we have already found: K (permeability), H (thickness) and Fluid Viscosity. Figure 12.85 is derived from analogs (oil fields in the GOM with API averaging around 24 API). Our fluid will have a higher API (32) and lower viscosity because we are being conservative.

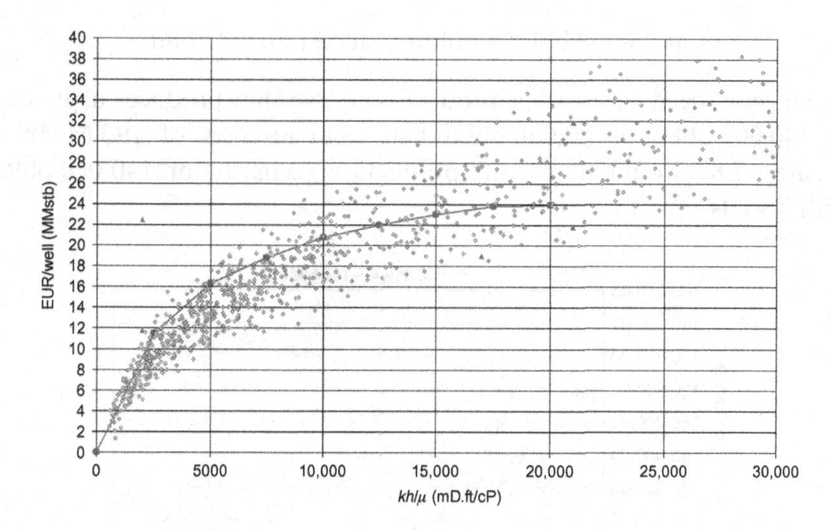

Figure 12.85 Estimation of number of wells based on rock and fluid properties.

Entering with KH equaling 100,000, Qi is about 7500 bbl/d. KH equals a 200,000 rate and could be up to 12,000 bbl/d or higher. Let's use the average 10,000 bbl/d. For a fluid of 32 API, these rates could be as high as 20,000 bbl/d.

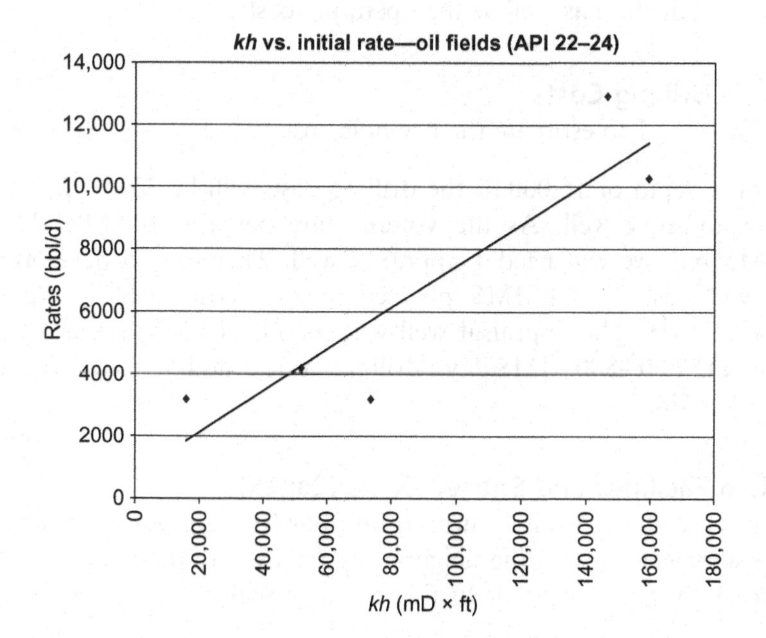

Figure 12.86 Estimation of well rates based on rock and fluid properties.

So 7 wells at 20,000 bbl/d would produce 140,000 bbl/d.

We now need to create a production curve that produces a plateau of 140,000 bbl/d or 60,000 bbl/d and total reserves of 190 MMbbl. Figure 12.87 would fit as our production scenario for 140,000 bbl/d with 7 wells.

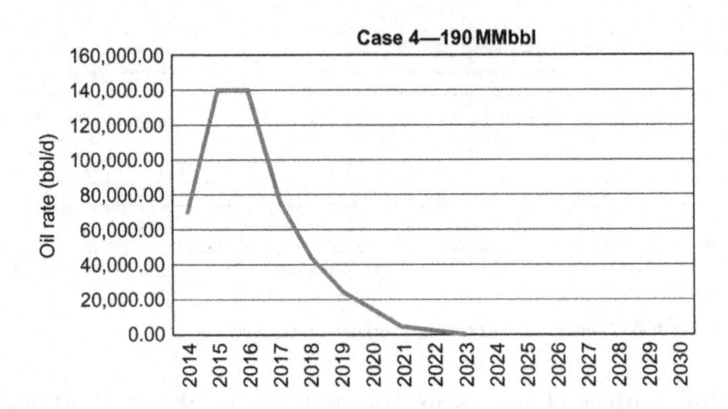

Figure 12.87 Production forecasting.

12.12.4 Costs: Drillex, Capex and Opex

Now it is necessary to estimate the cost to drill the 7 wells and evacuate the production as well as the operating costs.

12.12.5 Drilling Costs

Use Table 12.1 to estimate the dry hole cost.

For a depth of 15,000 ft, the drilling costs will be 35–64 MM$ for the exploratory well. As the volumes are between 100 MMbbl and 200 MMbbl, we will need 1 appraisal well. Therefore, a development well will cost 35–64 MM$ plus completion costs (+80%) equaling 63–115 MM$. The appraisal well will cost 35–64 MM$ plus 15 days more at 500–800 MM$/day drilling rate, adding up to equal 42.5–76 MM$.

12.12.6 Facilities and Subsea Costs (Capex)

In this case study, we may need a hub but let's explore the possibility for a subsea tieback to the neighboring area where there are already 5 prospects being developed. In this case, we will only need subsea and pipeline costs. For 7 wells, we will need only one manifold plus export pipeline to the platform and the riser. Costs should look like:

- Well templates and manifolds = 34 MM$/well
- SCR Production Riser = at 0.35 MM$/(″ × 1000 ft)
- Export Pipelines at 0.16 MM$/ (″ × miles). Let's assume that the export pipeline length = 5 miles

Templates and manifolds are 34 × 7 wells equaling 238 MM$, and production risers are 1 × 1 × water depth (i.e. 1.5 × 5500 ft) × 10″ × 0.35/1000 equaling 28.9 MM$. Export pipelines are 0.16 × 5 miles × 8″ equaling 6.40 MM$. As in this case, we have 140,000 bbl/d and may need a bigger diameter. So our riser may go up to 5 MM$ and our riser up to 2 MM$.

12.12.7 Operating Costs (Opex)

We will use the value for tieback provided by Table 12.2 with a tariff of 8–13 $/boe.

So for this project (subsea tieback with tariff), we will estimate our total Opex for the field life as a minimum of 190 MMbbl × 8 and maximum of 190 MMbbl × 13 equaling 1520 MM$ and 2470 MM$ for the 3 years of production on plateau.

This total cost should be prorated proportionally to production for the remaining years. For the peak rates on plateau, our yearly Opex would be 50% × 1520/3 equaling 253 MM$ or 50% × 2470/3 equaling 411.66 MM$ and prorated for the remaining years.

Again, a minimum of 25% Opex is recommended for the years where the production is lower than 25% of peak rate.

12.12.8 Project Timing and Input to Economic Analysis

A subsea tieback will require 3 years from exploration to production on average. We can drill all 7 wells during this time (2 wells per year). Our project may start up production in 2015 assuming we drill our exploratory well in 2011 and the appraisal one year later in 2012.

Date	Oil Production rate	Gas Production rate	Exploration Drilling	Appraisal Drilling	Development Drilling	Capex	Capex	Capex	Abandonment	Opex	Comments
	stb/day	MMscf/day	MM$	MMS	MM$	MM$	MM$	MM$	MM$	MM$	
			min-max	min-max	min-max	Facility	Subsea	pipeline			
2010											
2011			35-64								1 Exploratory
2012				42.5-76	(63-115) × 1						1 Appraisal + 1 well
2013					(63-115) × 2		119	28.9			2 wells
2014					(63-115) × 2		119	6.4			2 wells
2015	70,000.00				(63-115) × 2					126-205	2 wells
2016	140,000.00									253-411	
2017	140,000.00									253-411	
2018	77,000.00									126-205	
2019	45,000.00									63-102	
2020	25,000.00									63-102	
2021	15,000.00									63-102	
2022	5,000.00									63-102	
2023	2,500.00									63-102	
2024	1,000.00									63-102	
2025											
2026											
2027											
2028											
2029											
2030											
2031											
2032											
Total			35-64	42.5-76	441-805		238	35.3			

Figure 12.88 Economic analysis inputs.

12.13 CASE 13—EXAMPLE OF AN OFFSHORE FIELD WITH MULTI LAYERS (STACKED RESERVOIRS)

Make an economical analysis for a gas prospect, with a volume in place of 500 bcf (billions of cubic feet) of dry gas, no dissolved oil, and situated in deepwater Gulf of Mexico at water depth of 6900 ft. It is Upper Pliocene/Lower Pliocene age and of a total depth of 17,000 ft and 17,500 ft with an area of 2000 acres. Net pay equals 200 ft and 150 ft.

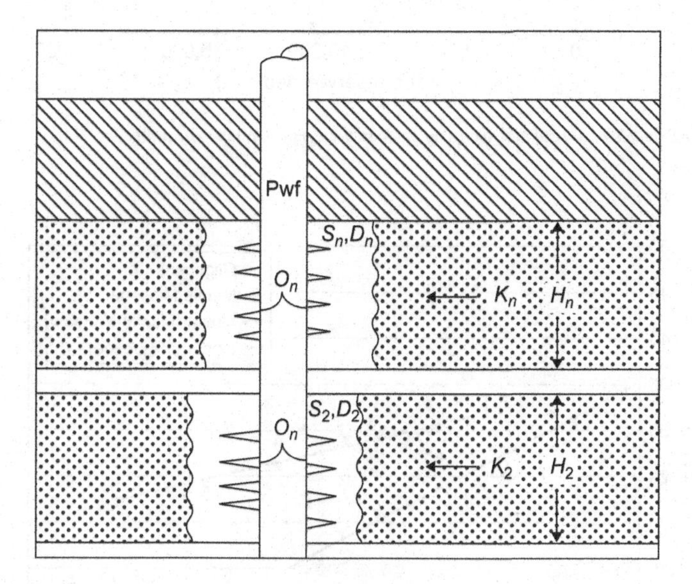

Figure 12.89 Schematic of a multilayer reservoir.

12.13.1 Recovery Factor Estimation

The first thing will be to estimate the recover factor for a gas accumulation. Figure 12.90 shows some RF analogs for gas fields (all Pliocene) in the GOM. We can see we should expect recoveries of 60% for depth of 17,000 ft. Therefore our recoverable volume will be 60% × 500 or 300 bcf.

12.13.2 Rock/Fluid Properties and Well Count Estimation

Next step is to derive the number of wells and their initial rate. To do this we first need to estimate some rock and fluid properties.

First, let's get an estimate for porosity. Looking at the dots for Miocene, we can see that for a depth of 17,000 ft porosity is around

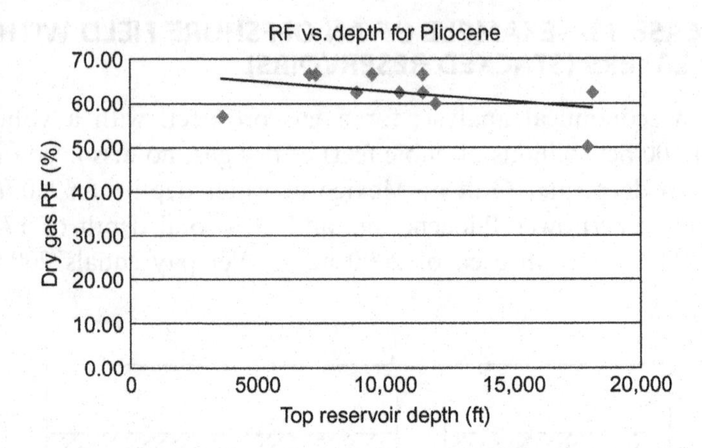

Figure 12.90 Example of a trend deriving using analogs, RF versus top reservoir depth.

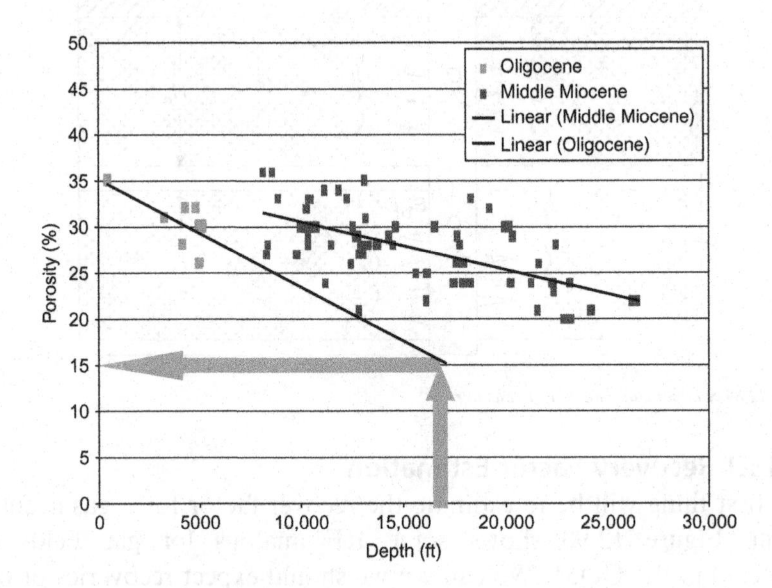

Figure 12.91 Trend for porosity vs. depth.

15%. Our porosity × permeability trend shows porosities starting at 20% giving a permeability range of <100 mD.

We have estimated so far recovery factor, porosity range and permeability ranges using analog information. We are assuming that the gas has a very low viscosity (i.e. <0.1 cP). Now that we have K and

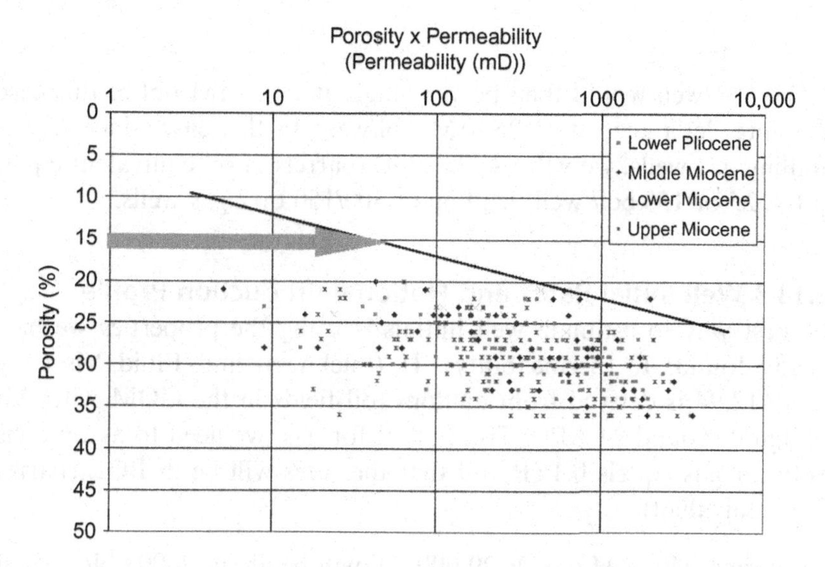

Figure 12.92 Porosity vs permeability trend.

H, we can estimate the recovery per well and therefore the total number of producer wells. K = 10–100 mD, H = 200 ft and Viscosity = 0.1 cP.

- Kh/mu (min) = 10 × 200/.1 = 20,000 mD ft/cP
- Kh/mu (max) = 100 × 200/.1 = 200,000 mD ft/cP

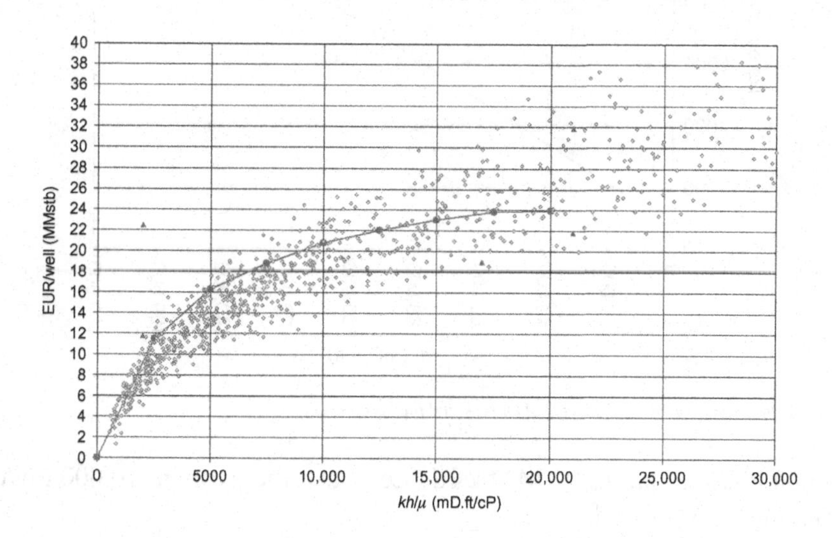

Figure 12.93 Estimation of number of wells based on rock and fluid properties.

The UR/well would then be very high at >25 MM bbl in this case. Therefore, let's keep it at 25 MM bbl/well. In this case, since we are handling gas wells, we will use 25 BOE (barrels of oil equivalent equaling 6 × 25 or 150 bcf/ well) leading to 300/150 or 2 gas wells.

12.13.3 Well Initial Rates and Notional Production Profile

The best way to estimate well initials is using the properties we have already found: K (permeability), H (thickness) and Fluid Viscosity. Figure 12.94 is derived from analogs (oil fields in the GOM with API averaging around 24 API). To apply it for gas, we need to assume viscosity for gas equals 0.1 cP and that the rates will be in BOE (barrels of oil equivalent).

Entering with KH equals 20,000, Qi will be about 2000 bbl/d equaling 12 MMscf/d. If KH equals a 200,000 rate, Qi could be up to

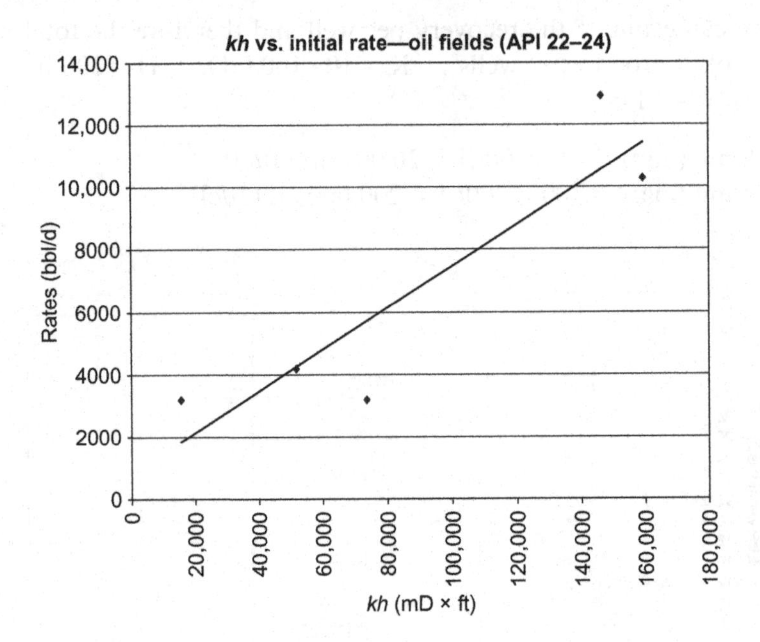

Figure 12.94 Estimation of well rates based on rock and fluid properties.

12,000 bbl/d equaling 72 MMscf/d. Let's use the average 10,000 bbl/d equaling 60 MMscf/d.

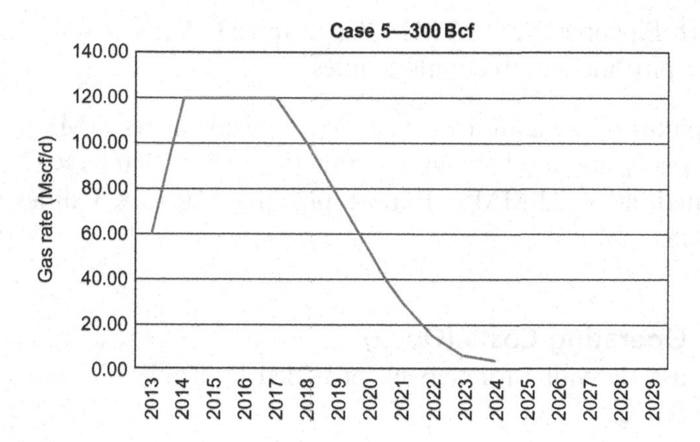

Figure 12.95 Production forecasting.

So with 2 wells at 60 MMscf/d, we would produce a plateau of 120 MMscf/d. Our production curve will look like that in Figure 12.95 for 120 MMscf/d with 2 wells.

12.13.4 Costs: Drillex, Capex and Opex

Now it is necessary to estimate the cost to drill the 2 wells and evacuate the production as well as the operating costs.

12.13.5 Drilling Costs

Use Table 12.1 to estimate the dry hole cost.

For a depth of 17,000 ft, the drilling costs will be 35–64 MM$ for the exploratory well. As the volumes are small, less than 100 MMbbl (300 bcf), we will not need any appraisal wells. Therefore, a development well will cost 35–64 MM$ plus completion costs (+ 80%) totaling 63–115 MM$.

12.13.6 Facilities and Subsea Costs (Capex)

In this case study, we may need a hub but let's explore the possibility for a subsea tieback to the neighboring area where there are already several prospects being developed. In this case, we will only need subsea and pipeline costs. For 2 wells, we will only need one manifold plus export pipeline to the platform and the riser. Costs should look like:

- Well templates and manifolds = 34 MM$/well
- SCR Production Riser = at 0.35 MM$/(″ × 1000 ft)

- Export Pipelines at 0.16 MM$/ (″ × miles). Let's assume that the export pipeline length equals 5 miles

Templates and manifolds equal 34 × 2 wells at 68 MM$, and production risers are 1 × 1 × water depth (i.e. 1.5 × 6900 ft) × 10″ × 0.35/1000 equaling 36.23 MM$. Export pipelines = 0.16 × 5 miles × 8″ or 6.40 MM$.

12.13.7 Operating Costs (Opex)

We will use the value for tieback provided by Table 12.2 with a tariff of 8–13 $/boe.

So for this project (subsea tieback with tariff), we will estimate our total Opex for the field life as a minimum of 50 MMbbl (300 bcf) × 8 and maximum of 50 MMbbl × 13 equaling 400 MM$ and 650 MM$ for the 4 years of production on plateau. This total cost should be prorated proportionally to production for the remaining years.

For the peak rates on plateau, our yearly Opex would be 50% × 400/4 equaling 50 MM$ or 50% × 640/4 equaling 80 MM$ and prorated for the remaining years. Again, the Opex distribution during the years is merely indicative and should be based on real operating costs for current and previous projects.

12.13.8 Project Timing and Input to Economic Analysis

A subsea tieback will require 3 years from exploration to production on average. We can drill the 2 wells in 2011. Our project may start up production in 2013 assuming we have already drilled our exploratory well in 2010.

Date	Oil Production rate	Gas Production rate	Exploration Drilling	Appraisal Drilling	Development Drilling	Capex	Capex	Capex	Abandonment	Opex	Comments
	stb/day	MMscf/day	MM$	MMS	MM$	MM$	MM$	MM$	MM$	MM$	
			min–max	min–max	min–max	Facility	Subsea	pipeline			
2010			35–64								1 Exploratory
2011					(63–115) × 2		34	36.23			2 wells
2012							34	6.4			
2013		60.00								25–40	
2014		120.00								50–80	
2015		120.00								50–80	
2016		120.00								50–80	
2017		120.00								50–80	
2018		100.00								25–40	
2019		75.00								25–40	
2020		51.00								25–40	
2021		30.00								12.5–20	
2022		15.00								12.5–20	
2023		6.00								12.5–20	
2024		3.00								12.5–20	
2025											
2026											
2027											
2028											
2029											
2030											
2031											
2032											
Total			35–64		126–230		68	42.63			

Figure 12.96 Economic analysis inputs.

12.14 CASE 14—EXAMPLE OF A FIELD OPTIMIZATION WITH HORIZONTAL WELLS

This example will be similar to case study 5, and the only difference is that we will now use horizontal wells instead of vertical wells. All other parameters will be the same.

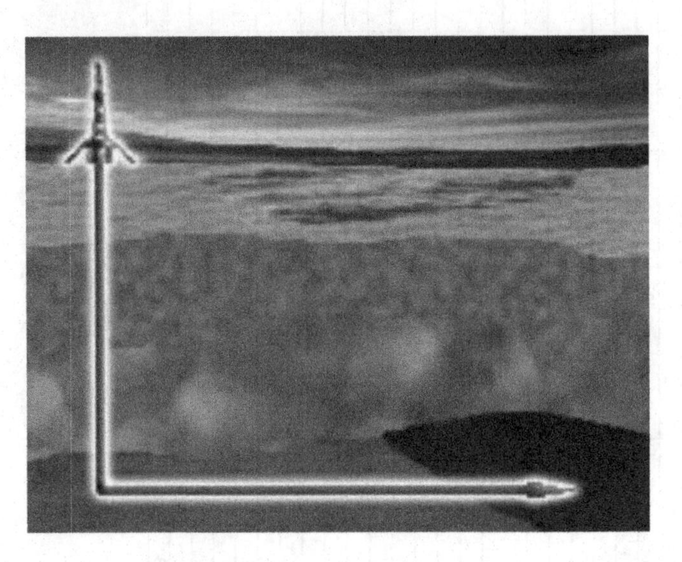

Figure 12.97 Drawing showing the horizontal well concept.

The horizontal wells will produce more and recover more than the vertical wells. However, they will cost more because to drill the horizontal length of 500 ft or more will require extra time.

12.14.1 Recovery Factor Estimation and Initial Rate

The first thing will be to estimate the recovery factor and initial rates for horizontal wells of different horizontal lengths. Figure 12.98 shows the modeling of a horizontal well of 500 ft. The initial rate can be 4 times higher than the one from a vertical well.

12.14.2 Rock/Fluid Properties and Well Count Estimation

Next step is to derive the number of wells and their initial rate. To do this, we first need to estimate some rock and fluid properties.

First, let's get an estimate for porosity. Looking at the dots for Miocene (Figure 12.99), we can see that for a depth of 17,000 ft,

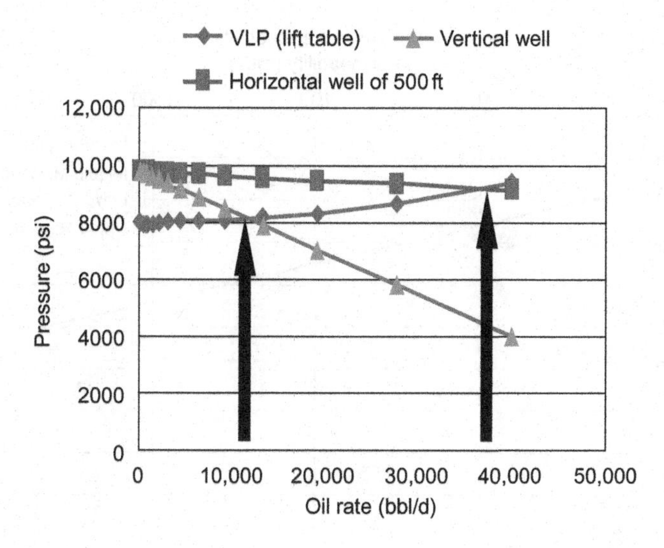

Figure 12.98 Inflow production for a vertical and a horizontal well.

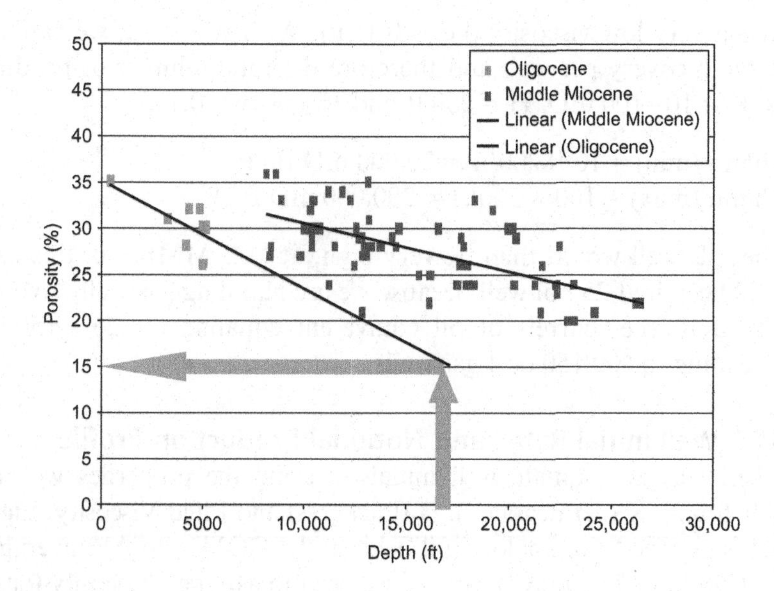

Figure 12.99 Trend for porosity vs. depth.

porosity is around 15%. Our porosity × permeability trends show porosities starting at 20%, which gives a permeability range of <100 mD.

We have estimated so far recovery factor, porosity range and permeability ranges using analog information. We are assuming that the

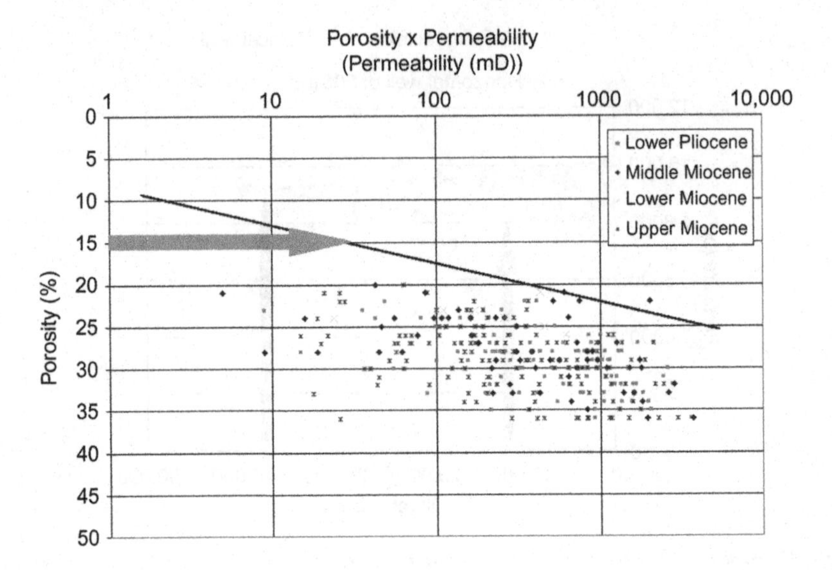

Figure 12.100 Porosity versus permeability trend.

gas has a very low viscosity (i.e. <0.1 cP). With K and H, we can estimate the recovery per well and therefore the total number of producer wells. K = 10–100 mD, H = 200 ft and Viscosity = 0.1 cP.

- Kh/mu (min) = 10 × 200/.1 = 20,000 mD ft/cP
- Kh/mu (max) = 100 × 200/.1 = 200,000 mD ft/cP

The UR/well would then be very high, at >25 MMbbl in this case. We will keep it at 25 bbl/well because we are handling gas wells. We will use 25 BOE (i.e. barrels of oil equivalent equaling 6 × 25 = 150 bcf/well) leading to 600/150 or 4 gas wells.

12.14.3 Well Initial Rates and Notional Production Profile

The best way to estimate well initials is using the properties we have already found: K (permeability), H (thickness) and Fluid Viscosity. Figure 12.101 is derived from analogs (oil fields in the GOM with API averaging around 24 API). To apply it for gas, we need to assume a viscosity for gas equaling 0.1 cP and that the rates will be in BOE (barrels of oil equivalent).

Entering KH = 20,000, Qi is about 2000 bbl/d equaling 12 MMscf/d. For KH equaling a 200,000 rate, the Qi could go up to 12,000 bbl/d equaling 72 MMscf/d. Let's use the average 10,000 bbl/d equaling 60 MMscf/d.

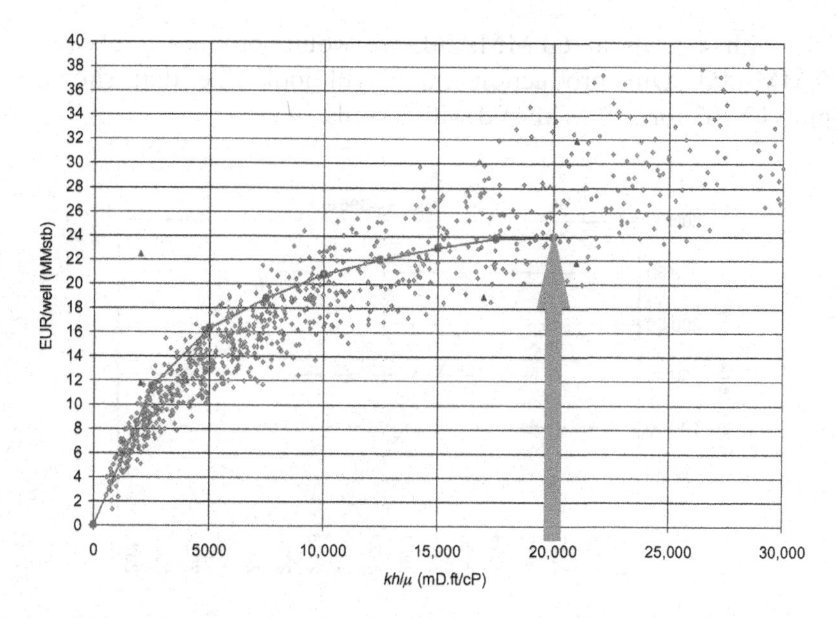

Figure 12.101 Estimation of number of wells based on rock and fluid properties.

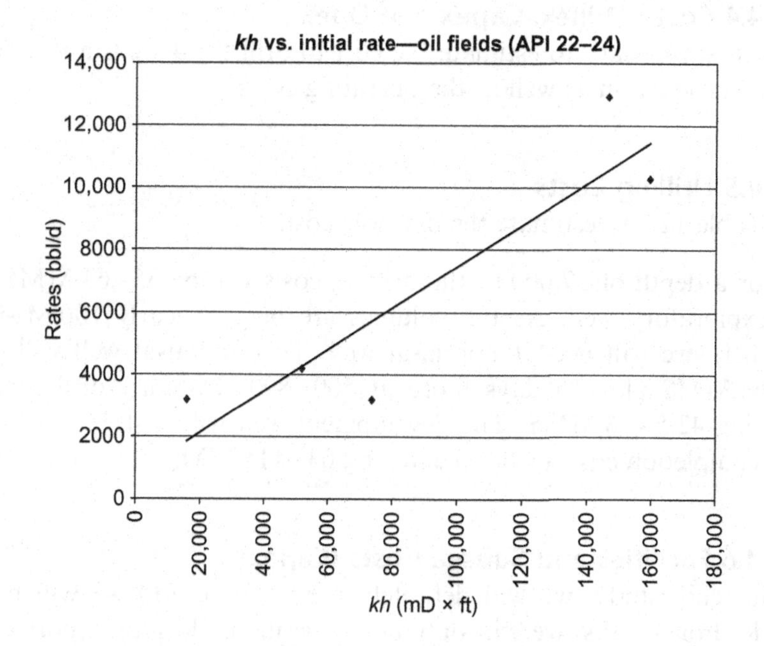

Figure 12.102 Estimation of well rates based on rock and fluid properties.

So with 4 wells at 60 MMscf/d, we would produce a plateau of 240 MMscf/d. Our production curve will look like that shown in Figure 12.103 for 240 MMscf/d with 4 wells.

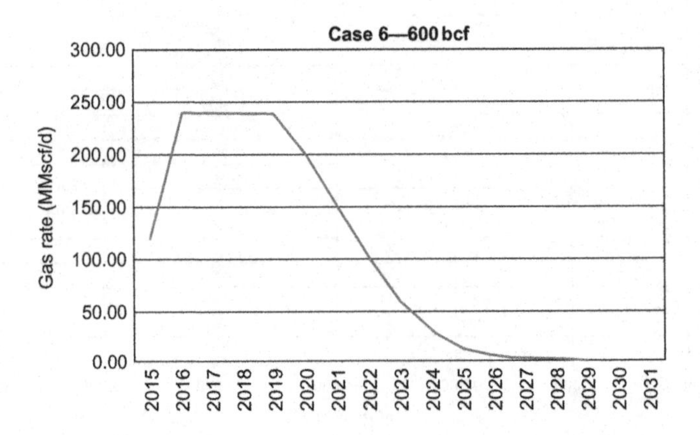

Figure 12.103 Production forecasting.

12.14.4 Costs: Drillex, Capex and Opex

Now it is necessary to estimate the cost to drill the 4 wells and evacuate the production as well as the operating costs.

12.14.5 Drilling Costs

Use Table 12.1 to estimate the dry hole cost.

For a depth of 17,000 ft, the drilling costs will be 35–64 MM$ for the exploratory well. As the volumes are bigger, nearly 100 MMbbl (600 bcf), we will need 1 appraisal well. This appraisal well will cost 35–64 MM$ plus 15 days more at 500–800 MM$/day drilling rate equaling 42.5–76 MM$. The development well will cost 35–64 MM$ plus completion costs (+ 80%) equaling 63–115 MM$.

12.14.6 Facilities and Subsea Costs (Capex)

In this case study, we will definitely need a hub, and we will use a SPAR. For 4 wells, we will only need one manifold plus export pipeline from the SPAR to an existing infrastructure (100 miles and 25″ diameter).

Figure 12.104 will be used again to provide an estimation of Capex for a SPAR of capacity 250 MMscf/d (~50 kbbl/d) equaling 400 MM$ for Hull.

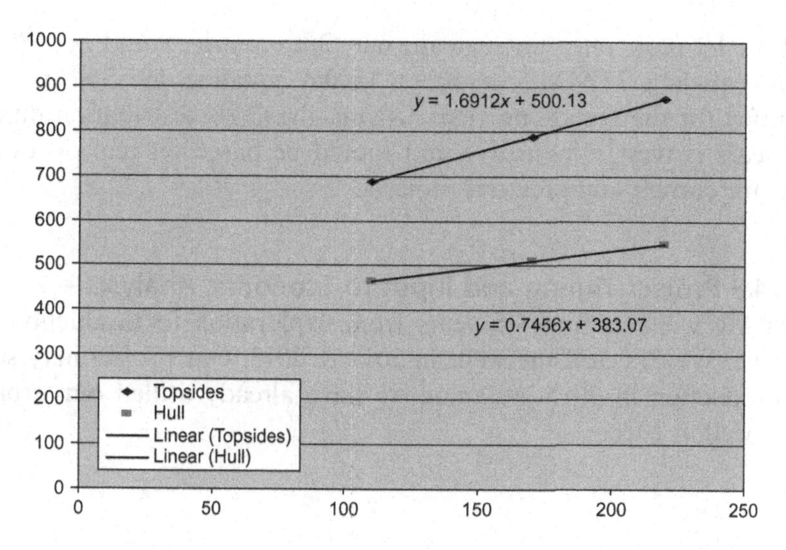

Figure 12.104 Hull and topsides Capex estimation MM$ versus FPU capacity.

And for the subsea, costs should be:

- Well templates and manifolds = 34 MM$/well
- SCR Production Riser (diameter = 8″) = at 0.35 MM$/(″ × 1000 ft)
- Export Pipelines at 0.16 MM$/ (″ × miles). Let's assume that the export pipeline length equals 100 miles and 25″ diameter
- Intra-field flow lines at 0.22 MM$/(″ × miles) at 8 inches and 5 miles

Templates and manifolds are 34 × 4 wells equaling 136 MM$, and production risers are 1 × 1 × water depth (i.e. 1.5 × 6900 ft) × 8″ × 0.35/1000 equaling 28.98 MM$. Export pipelines are 0.16 × 100 miles × 25″ equaling 400 MM$, and intra-field flow lines are 0.22 × 5 miles × 8 equaling 8.8 MM$.

12.14.7 Operating Costs (Opex)

We will use the value for Standalone FPU (SPAR) provided by Table 12.2 with a no tariff 8–15 $/boe as we will take care of building our own pipeline for the gas evacuation.

So for this project, we will estimate our total Opex for field life as a minimum of 100 MMbbl (600 bcf) × 8 and maximum of 100 MMbbl × 15 equaling 1500 MM$. This total cost should be prorated proportionally to production for the remaining years.

For the peak rates on plateau, our Opex yearly would be 50% × 600/4 equaling 75 MM$ or 50% × 1500/4 equaling 187.5 MM$ and prorated for the remaining years. Again, the Opex distribution during the years is merely indicative and should be based on real operating costs for current and previous projects.

12.14.8 Project Timing and Input to Economic Analysis

A SPAR will require 4–5 years from exploration to production on average. We can drill the wells in 2012 & 2013. Our project may start up production in 2015 assuming we have already drilled our exploratory well in 2010.

Date	Oil Production rate	Gas Production rate	Exploration Drilling	Appraisal Drilling	Development Drilling	Capex	Capex	Capex	Abandonment	Opex	Comments
	stb/day	MMscf/day	MM$	MMS	MM$	MM$	MM$	MM$	MM$	MM$	
			min-max	min-max	min-max	Facility	Subsea	pipeline			
2010			35-64								1 Exploratory
2011				42.6-76			34				1 Appraisal
2012					(63-115) × 2	200	34	8.8			2 well
2013					(63-115) × 2	200	34	28.98			2 well
2014							34	400			
2015		120.00								37.5–93.75	
2016		240.00								75–187.5	
2017		240.00								75–187.5	
2018		240.00								75–187.5	
2019		240.00								75–187.5	
2020		200.00								75–187.5	
2021		150.00								37.5–93.75	
2022		102.00								37.5–93.75	
2023		60.00								18.25–46.875	
2024		30.00								9.12–23.4	
2025		12.00								9.12–23.4	
2026		6.00								9.12–23.4	
2027		3.00								9.12–23.4	
2028		2.00								9.12–23.4	
2029		1.00								9.12–23.4	
2030										9.12–23.4	
2031											
2032											
Total			35–64	42.6–76	252–460	400	136	437.78			

Figure 12.105 Case 6—Economic analysis inputs.

12.15 CASE 15—EXAMPLE OF A FIELD UNITIZATION CASE

This case study will cover a group of fields which belong to different partnerships. The partners involved will want to have a common interest among them in all the blocks. This can be done only if they unify by selling or acquiring interest according to their volumes owned and share what they have in each of the blocks.

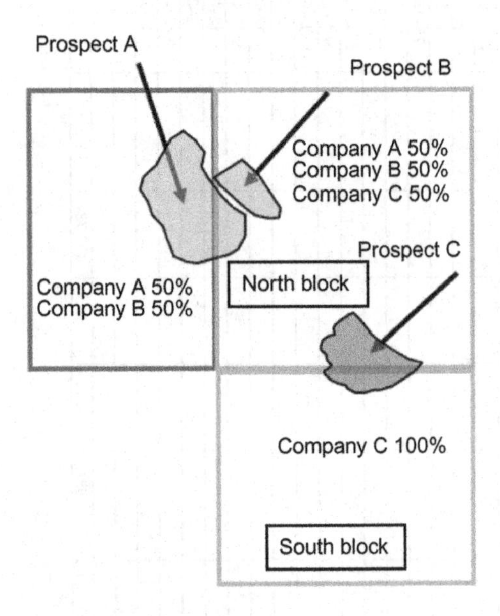

Figure 12.106 Location of a group of prospect to be unitized.

In this particular case study, we will concentrate on the unitization. Therefore, let's assume that the recoverable volumes are known or have been estimated by the methods described in the previous examples.

12.15.1 Calculation of the Individual Shares Based on Volume

In this case study, there are 3 companies with different shares on each of the 3 blocks. Prospect A is 200 MMbbl, Prospect B is 100 MMbbl and Prospect C is 150 MMbbl. The shares for each company are indicated in Figure 12.106. The objective is first to get an agreement on prospect C which is 30% in the south block and the remainder in the north block. Therefore, the corresponding volumes for prospect C will be 45 MMbbl in the south and 90 MMbbl in the north.

Table 12.4 shows that the new share will be based on recoverable volume and the percentage on the volume in each block. Total volume is 45 in the south plus 90 in the north equaling 135 MMbbl. Companies A and B have no share in the South (= 0) and Company C has 100% in the South and 25% in the north. As it can be seen, Company C retain 50%, B 33.33 % and A with 16.67%, from the initial arrangement of 25%, 50% and 25% respectively.

Table 12.4 Table to Obtain Shares based on Volumes

	Volumes			**Volumes**				
	North Block		Shares	South Block		Shares		%
	90		1	45		1	135	
Company A	90		0.25	45		0	22.5	16.67%
Company B	90		0.5	45		0	45	33.33%
Company C	90		0.25	45		1	67.5	50.00%

12.15.2 Calculation of the Individual Shares Based on Production

Let us assume now that we are not sure about the volume percentages between south and north and that our development plan assumes all the wells will be drilled only in the north block.

The production that will be coming from the south (estimated based on reservoir simulation) will be only 20%, although 30% of the volume may be in the south, the wells will drain more from the north than from the south.

This is possible if the reservoir quality is better in the north as one example (Table 12.5).

Table 12.5 Table to obtain Share based on Production or Drainage

	Production	**80% of 135**		**Production**	**20% of 135**			
	North Block		Shares	South Block		Shares		%
	108		1	27		1	135	
Company A	108		0.25	27		0	27	20.00%
Company B	108		0.5	*21*		0	54	40.00%
Company C	108		0.25	27		1	54	40.00%

As we can see, the production scheme and the development plan will benefit the companies with shares in the north because the reservoir has better economic value there, and therefore, the north is preferred.

The same methodology can be applied now to prospect A in the north/northwest block.

BIBLIOGRAPHY

Allen, J.R., The petroleum play. In: Basin Analysis, Blackwell Science Ltd., 1990, pp. 309–396.

Amado, L., et al., Integrated Prospect Evaluation Using Electromagnetic, Seismic, Petrophysical, Basin Modeling, and Reservoir Engineering Data, AAPG Houston, 2006.

Amado, L., et al., Evaluation of Reservoir Modeling Techniques to Assess the Impact of Stratigraphic Architecture on Production Performance, APPG Houston, 2006.

Amado, L., Enabling Technology, An overview of Offshore Innovations to Unlock Difficult Plays, Talk given at 2009 EP Global Summit, Barcelona, Spain, January 2009.

Amado, L., Enabling Technology in Deep Water Reservoirs, Poster session presented at Rio Offshore Conference in Rio de Janeiro Brazil, September 2008, IBP—Brazilian Petroleum Institute.

Amado, L., Enabling Technology, An Overview of Offshore Innovations in the Oil Industry, SPE Paper 120919 to be presented at LACPEC 2009, Cartagena, Colombia, June 2009.

Amado, L., EP Global Summit in Barcelona Spain, Paper presented: "Pushing the Limits for Deepwater Offshore Exploration and Production in the GOM", 2009.

Dake, L.P. and Towler, B., Fundamentals of Reservoir Engineering, 2nd Edition, 2011.

E&P Facilities Decommissioning Techniques, Overview of Offshore Facilities.

IHS—Oil and Gas Information and Analytical Tools <www.ihs.com>.

Jahn, F., Cook, M., Graham, M., Hydrocarbon Exploration and Production. Developments in Petroleum Science, Elsevier, 2008.

McCain, W. D. Jr., The Properties of Petroleum Fluids. Tulsa, OK; PennWell Publishing Company, 1990.

Gulf Professional Publishing is an imprint of Elsevier
The Boulevard, Langford Lane, Kidlington, Oxford, OX5 1GB, UK
225 Wyman Street, Waltham, MA 02451, USA

First published 2013

Notices
Knowledge and best practice in this field are constantly changing. As new research and experience broaden our understanding, changes in research methods, professional practices, or medical treatment may become necessary.

Practitioners and researchers must always rely on their own experience and knowledge in evaluating and using any information, methods, compounds, or experiments described herein. In using such information or methods they should be mindful of their own safety and the safety of others, including parties for whom they have a professional responsibility.

To the fullest extent of the law, neither the Publisher nor the authors, contributors, or editors, assume any liability for any injury and/or damage to persons or property as a matter of products liability, negligence or otherwise, or from any use or operation of any methods, products, instructions, or ideas contained in the material herein.

British Library Cataloguing in Publication Data
A catalogue record for this book is available from the British Library

Library of Congress Cataloging-in-Publication Data
A catalog record for this book is available from the Library of Congress

ISBN: 978-1-85617-853-2

DEDICATION

I would like to dedicate this book to the following people:

To my family: My wife Vania and my son Liam, who have always supported me in doing this project and motivated me to finish it.

To my parents, for the example given to me to always look forward.

To my colleagues and coworkers in the oil companies I worked for and those who have provided valuable feedback to me along my career.

Printed in the United States
By Bookmasters